# TI-83 PLUS® COMPANION

Larry A. Morgan
*Montgomery County Community College*

*to accompany*

# Elementary Statistics

## Eighth Edition

Mario F. Triola

**Boston San Francisco New York**
**London Toronto Sydney Tokyo Singapore Madrid**
**Mexico City Munich Paris Cape Town Hong Kong Montreal**

To Abui, Shaine, Cheyenne, and Miles

The programs and applications presented in this book have been included for their instructional value. They have been tested with care but are not guaranteed. The publisher does not offer any warranties or representations, nor does it accept any liabilities with respect to the programs or applications.

Many of the designations used by manufacturers and sellers to distinguish their products are claimed as trademarks. Where those designations appear in this book, and Addison Wesley Longman was aware of a trademark claim, the designations have been printed in initial capital letters or all capital letters.

Reproduced by Addison-Wesley Publishing Company Inc. from camera-ready copy supplied by the authors.



ISBN 0-201-70469-2

2 3 4 5 6 7 8 9 10 PHTH 03 02 01

# Preface

The TI-83 Plus is a little computer with many capabilities. This companion will introduce you to some of its statistical capabilities as they relate to *Elementary Statistics* (8th ed.) by Mario F. Triola. (The TI-83 works like the TI-83 Plus for most of the procedures in this companion, differences will be pointed out.) This companion is designed to be used side by side with the text. For example, Chapter 1 of *Elementary Statistics* has a definition of a *simple random sample* [pg. 19] and Exercise 17 [pg. 24] requests such a sample. Chapter 1 of this companion shows how to select a random sample with the TI-83 Plus (on page 4) and refers to the main text by having the main text page numbers in square brackets.

This companion follows the text of *Elementary Statistics* chapter by chapter, offering helpful techniques on the TI-83 Plus. Most of the examples and data are from the text; these examples are paraphrased but should be complete enough for continuity. Consult the text for important details such as the appropriateness of the procedure used.

Chapter 1, and Chapter 2 are very important for your understanding of all the chapters that follow including the TI-83 Plus keyboard notation. Know the meanings of such phrases as "Home screen" and "the last entry feature," and the ways to perform such procedures as saving and deleting data. You will need to understand these terms and procedures before you use the material in later chapters.

Even though the TI-83 Plus can often perform operations in more than one way, I do not show all the possibilities but concentrate instead on the method that will be most valuable in the long run.

Real data sets are used throughout the main text and provided in printed form [in Appendix B] but can be installed in your TI-83 Plus from another TI-83 Plus or from a computer and are available from the CD-ROM that came with your main text. See the appendix of this companion for more details.

The back inside cover of this companion has a "TI-83 Plus Quick Reference" that summarizes some important keys, menus, and functions used in this companion.

Although this is a small book, I had a large amount of help with it. I thank my students and colleagues who have supported the use of graphing calculators in statistics classes, with special thanks to Roseanne Hofmann. Thanks to Charlotte Andreini and Michelle Miller of Texas Instruments. Thanks, of course, to Mario F. Triola, who made this companion possible. Thanks to Deirdre Lynch, Rebecca Martin, Cindy Cody, and the staff of Addison Wesley Longman Inc. for their assistance.

# Contents

| KEY LOOKUP TABLE | |
|---|---|
| Key | At |
| 2nd | A2 |
| ALPHA | A3 |
| ↑ ANS | D10 |
| APPS | B4 |
| CLEAR | E4 |
| DEL | C2 |
| ↑ DISTR | D4 |
| ENTER | E10 |
| ↑ ENTRY | E10 |
| GRAPH | E1 |
| ↑ INS | C2 |
| ↑ LIST | C3 |
| MATH | A4 |
| ↑ MATRX | A5 * |
| ↑ MEM | E9 |
| MODE | B2 |
| ON | A10 |
| PRGM | C4 |
| ↑ QUIT | B2 |
| ↑ RCL | A9 |
| STAT | C3 |
| ↑ STATPLOT | A1 |
| STO▸ | A9 |
| TRACE | D1 |
| VARS | D4 |
| WINDOW | B1 |
| Y= | A1 |
| ZOOM | C1 |
| ÷ for / | E6 |
| x for * | E7 |

* B4 on the TI-83

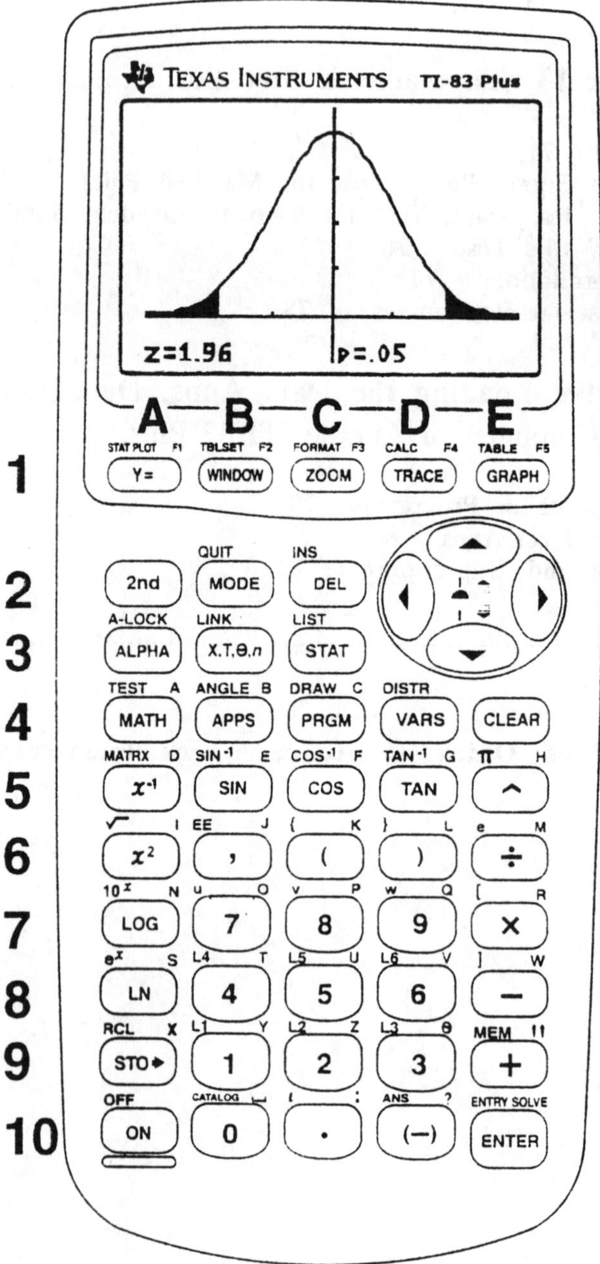

# 1 Introduction to Statistics (and to the TI-83 Plus)

In this chapter we introduce our calculator companion to Triola's *Elementary Statistics* (8th ed.) by giving an overview of the TI-83 Plus keyboard and give the notation used for the various keys and menus. Read this section carefully so you will know which keys are being referred to throughout this manual.

You will also learn how to set the correct MODE on the TI-83 Plus to ensure that you will obtain the same results as this companion does. And so you do not end up squinting at the screen trying to read it as it gets ever dimmer, you will find out how to adjust the screen contrast to keep it sharp and how to check the battery strength.

In this chapter we begin our study of statistics with Triola's *Elementary Statistics* (8th ed.) by taking an exercise from the text and showing you how to select a simple random sample. The example also introduces the notation that will be used throughout this companion when menu items are selected. You will then learn how to do basic calculations on the Home screen. The Home screen keeps past entries in sight so that you can edit or modify them and then store them for future use. You will then see how lists of data can be saved and edited in the STAT editor. You will need the above skills throughout this manual so work along with your calculator and refer to this material when needed.

In this chapter, as in the chapters that follow, we do not summarize all the important text information. Our purpose in this companion is to help you use the TI-83 Plus in conjunction with the text. Be sure to study the text for a full discussion of definitions, concepts, and proper statistical procedures.

## KEYBOARD AND NOTATION

The TI-83 Plus keyboard has 5 columns (designated in the schematic, shown on the opposite page, as A, B, C, D, E) and 10 rows of keys. The cursor control, or arrow, keys ( ▲ ▼ ◀ ▶ ) toward the upper right of the keyboard disturb the pattern but in a logical way.

Start in the upper left corner with the **Y=** key in the A column and the first row, or the A1 location. (Touch the keys mentioned as you follow along.) The **ON** key is at A10, **ENTER** is at E10, and the **GRAPH** key is at E1. These four keys are the corners of the TI-83 Plus keyboard.

The alphabetized Key Lookup Table, to the left of the schematic, will be helpful in finding keys mentioned in this companion. After each key name is its location on the keyboard.

Many keys on the TI-83 Plus have multiple functions. The primary function is marked on the key itself, and other functions are marked in colors above the key. Let's see how these functions marked in colors work.

### Yellow 2nd Key at A2

The **STAT** key is at C3; above it, in yellow, is **↑LIST**, which is engaged by pressing and releasing the yellow **2nd** key and then the **STAT** key. (You know the **2nd** key has been engaged when the cursor key is a blinking up arrow.) Other such key combinations will be for **↑QUIT**, **↑INS**, **↑ANS**, and **↑ENTRY**.

In this manual a small up arrow before a word designates that the word is in yellow and thus you must hit the yellow **2nd** key first. You must also press the **2nd** key first with the yellow list designations **L1**, **L2**, . . . , **L6** above the numerals **1**, **2**, . . . , **6** and with the yellow { } set, or list braces, above the regular parenthesis keys ( ) at C6 and D6. No up arrow is used for these, however, because their locations are easy to remember.

### Green ALPHA Key at A3

You can use the letters of the alphabet to assign data storage locations and to name a list. To engage them, first press and release the **ALPHA** key. For example, the letter **G** is engaged by hitting **ALPHA** and then pressing the **TAN** key at D5, which has the letter **G** above it to the right in green. You know the **ALPHA** key is engaged when the cursor is a blinking **A**.

### Some General Keyboard Patterns and Important Keys

1. The first row is for plotting and graphing.
2. The second row has the important **2nd ↑QUIT** combination, the **DEL ↑INS** key, and the cursor control keys, which are all used for editing. Just below the cursor control keys is the **CLEAR** key at E4.
3. The A column has the **MATH** key and various math functions, for example, $x^2$, √.
4. The E column has the math operations +, −, ×, ÷, ^.
   **Note**: The ÷ shows on the TI-83 Plus screen as / and the × as *.
5. The **STAT ↑LIST** key at C3 and the nearby **VARS ↑DISTR** key at D4 will be central to our study of statistics and important probability distributions.
6. The sixth row, above the numeral rows, has ( { , } ), which are used for grouping and spacing.
7. The **STO▸** key at A9 is used for storing. It shows as a → on the display screen.
8. The tenth row has the negative symbol **(-) ↑ANS** key, which differs from the subtraction key at E8, and the biggest key, the **ENTER ↑ENTRY** key.
   **Note**: The (-) shows as ⁻ on the screen—smaller and higher than the subtraction sign.

## SETTING THE CORRECT MODE AT B2

If your answers do not show as many decimal places as the examples in this companion have, or if you have difficulty matching other output, check your **MODE** settings.

From the Home screen, the screen that appears when you turn on your TI-83 Plus, press **MODE** at B2. The menu on the right appears, with the first word in each row darkened. If your screen looks different, use the cursor control key to go to each divergent row and then, with the first element blinking, press **ENTER**. Repeat this procedure until the screen looks like the one at the right.

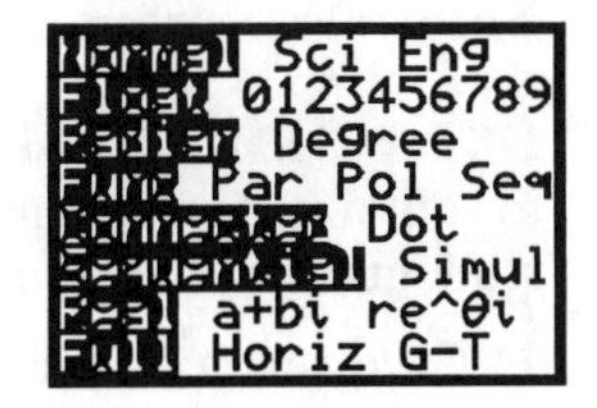


Engage ↑**QUIT** at B2 to return to the Home screen.

**Note**: All other settings are assumed to be factory or default settings and may be restored as explained on pages 18-6 and 18-7 of the Guidebook that comes with the TI-83 Plus.

## SCREEN CONTRAST ADJUSTMENT AND BATTERY CHECK

To adjust the screen contrast, follow these steps:

From the Home screen, press and release the **2nd** key and hold down the ▲ cursor control key to increase the contrast. Notice that the number in the upper right corner of the screen increases from 0 (lightest) to 9 (darkest). Press and release the **2nd** key and hold down the ▼ key to lighten the contrast. (A visual reminder of this appears on the TI-83 Plus with the yellow and black circle between the cursor keys on the keypad.) If you adjust the contrast setting to 0, the display may become completely blank. Press and release the **2nd** key and then hold down the ▲ key until the display reappears.

When the batteries are low, the display begins to dim (especially during calculations), and you must adjust to a higher contrast setting. If you have to set the contrast setting to 9, you will need to replace the four AAA batteries soon. (If your batteries are low, the TI-83 Plus displays a low-battery message when you turn on the calculator.) The display contrast may appear very dark after you change batteries. Press and release the **2nd** key and then hold down the ▼ key to lighten the display.

**Note**: The Important!! message pasted on the battery cover of the TI-83 Plus instructs to turn off your calculator immediately if you see any low battery message and replace the batteries and perform garbage collection (as explained in the TI-83 Plus Guidebook). Failure to immediately complete the steps above may result in loss of your data and corruption of (calculator) memory.

## RANDOM SAMPLES AND MENU NOTATION

**DEFINITION** [pg. 19]: A *simple random sample* of size *n* subjects is selected in such a way that every possible sample of size *n* has the same chance of being chosen.

**Note**: The numbers in brackets refers to the pages in *Elementary Statistics*.

**EXERCISE 17 (modified)** [pg. 24]: Describe in detail a method that could be used to obtain: (a) a randomly selected student to 'volunteer' in a statistics class of 28 students. (b) a simple random sample of the heights of five of the students.

1. Give each student a different consecutive counting number starting from 1 as an ID number. For example, you would use 1 to 28 for a class of 28 students.

2. Press the **ON** key on the TI-83 Plus. A cursor should be blinking on the Home screen. If not, press **↑QUIT**, at B2, to return to the Home screen.

3. Press the **CLEAR** key if the cursor is not in the upper left corner.
   **Note**: **CLEAR**, at E4, clears a line to the left *or* the screen above it.

4. We will generate five random numbers from 1 to 28. The numbers are random in the sense that each has an equal chance of being generated; but they are generated mathematically so that you can get the same results as shown below by setting the same seed as follows.

   (a) On the Home screen, type **123**; then press **STO▸**, then **MATH**, then **◀**, the left cursor control key *once* (or the right cursor control key *three* times), to highlight PRB for the menu shown in screen (2).

   (b) Then press **1**, and **rand** is pasted to the Home screen, as shown in the first line in screen (3).

   (c) Press **ENTER**, and the 123 on the second line of screen (3) indicates that the seed is set.
   **Note**: In the future this type of sequence of steps will be given as 123 **STO▸ MATH** <PRB> **1**:rand or 123 **STO▸**rand.

```
MATH NUM CPX PRB
1:rand
2:nPr
3:nCr
4:!
5:randInt(
6:randNorm(
7:randBin(
```
(2)

5. Press **MATH** <PRB> **5**:randInt(, and then type **1,28** (be sure to type the comma); then press **ENTER** for the first line of screen (4) with 20 being the ID of the class 'volunteer'. Press **ENTER** six more times because of the repeats of 21 and 13 that occurred.

   **Note**: If we used randInt(**1,28,5** and pressed **ENTER** twice you would have the two lists of numbers in screen (5). If randInt(**1,28,10** were used, the list of numbers would not fit on one line and we could use the ▶ key to reveal the unseen numbers. More on lists on page 7.

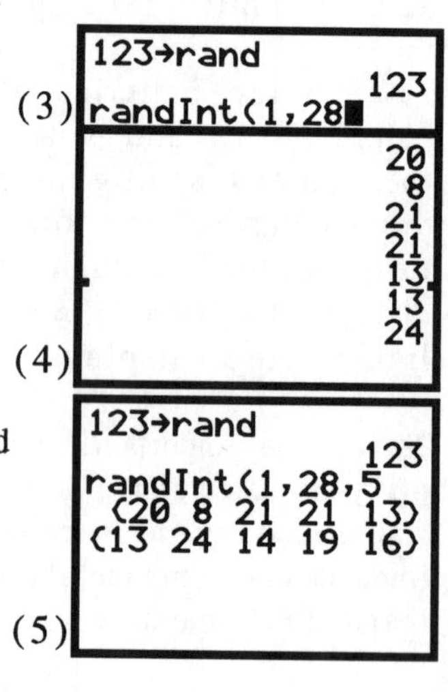


6. Measure the heights of the students with ID numbers 20, 8, 21, 13, and 24.

**EXAMPLE**: Use the above procedure to pick seven integers from 1 to 74 to play a state Lotto game.

One possibility is 43, 26, 58, 40, 14, 71, and 20, as shown in screen (6).

**Note**: There is no need to set the seed if you are not interested in duplicating the results.

```
4321→rand
                4321
randInt(1,74,5
{43 26 58 40 14}
{71 71 20 52 42}
```

(6)

## HOME SCREEN CALCULATIONS AND STORING RESULTS

**CUMULATIVE REVIEW EXERCISES** [pg. 28]: Calculator warm-ups.

We will use the first two exercises to show some Home screen techniques for dealing with data. Solutions for each of the other exercises (3 to 10) will be given so that you can get some practice with these techniques.

1. $$\frac{1.23 + 4.56 + 7.89}{3}$$

### The DEL ↑INS Key, Location 'Ans', and the 'Last Entry Feature'

(a) Type **1.23+4.56+1.789**. The last number (1.789) is an intentional mistake.

(b) Press ◀◀◀ (the left cursor control key three times) to place the cursor over the unwanted 7, as shown in screen (7). Press **DEL**, at C2, and the 7 is deleted. Press ◀◀ to move the cursor to the 1 before the decimal, and type **7**, overwriting the 1. Press **ENTER** for the answer of 13.68, the sum of the numerator, as shown in screen (8).

```
1.23+4.56+1.■89
```

(7)

```
1.23+4.56+7.89
               13.68
Ans/3
                4.56
```

(8)

(c) Press ÷ and 'Ans/' appears on the screen. Type **3** and press **ENTER** for the desired results of Exercise 1 of 4.56.

**Note**: 'Ans' represents the last value or results of a calculation displayed alone and right justified on the Home screen. Pressing ÷ without typing a value before it called for something to be divided, so 'Ans' was supplied.

(d) Press **↑ENTRY** to return the 'last entry' or 'Ans/3.' Press **↑ENTRY** again, and the numerator is returned to the screen.

(e) Press ▲ (the up cursor control key) to jump to the 1 at the front of the line).

(f) Press **↑INS**, at B3, to change the blinking rectangular cursor to the blinking underline of the insert mode. Type **(** , the left parenthesis at C6, to insert a left parenthesis before the numerator.

(g) Press ▼ to jump to the end of the line. Type **)÷3** , to insert a right parenthesis after the numerator and to divide by 3.

(h) Press **ENTER** for the same results as before (4.56), as shown in the bottom lines of screen (9).

```
1.23+4.56+7.89
           13.68
Ans/3
            4.56
(1.23+4.56+7.89)
/3
            4.56
```

(9)

2. $$\sqrt{\frac{(5-7)^2+(12-7)^2+(4-7)^2}{3-1}}$$

**↑ANS Key, Syntax Errors, and the ALPHA Key and Storage.**

(a) Type **(5−7)²+(12−7)²+(4−7)²** with **²** from the **x²** key at A6. Press **ENTER** for the value 38, as in the top part of the screen (10).

(b) Press **↑√** (above the **x²** key at A6). Press **↑ANS**, at D10; then press **÷(3−1))**, then **ENTER** for the desired results of 4.36, also shown in the last line of screen (10).

(c) Both (a) and (b) could have been done in one step. Screen (11) shows a first attempt. Pressing **ENTER** brings screen (12), which indicates a syntax error.

Press **2** to Goto the error, and the cursor will blink over the last parenthesis as in screen (13). This indicates one more right parenthesis than left, so we need to add another left parenthesis in the beginning, as shown in screen (14).

(d) We now want to store parts of the above problem in different locations. Type **(5−7)²**, and press **STO▸** at A9; then press **ALPHA**, at A3, then **R** at E7, then press **ENTER** for '4' as on top of screen (15). Repeat typing **(12−7)²** and **(4−7)²** and storing in **S** and **T**. Find the sum and store the result in **H** by typing **R+S+T STO▸ H**. for 38 as shown in the top lines of screen (16) You can then use the sum for the final results by typing **↑√ ALPHA H ÷(3−1))**, then **ENTER** for the last lines of screen (16).

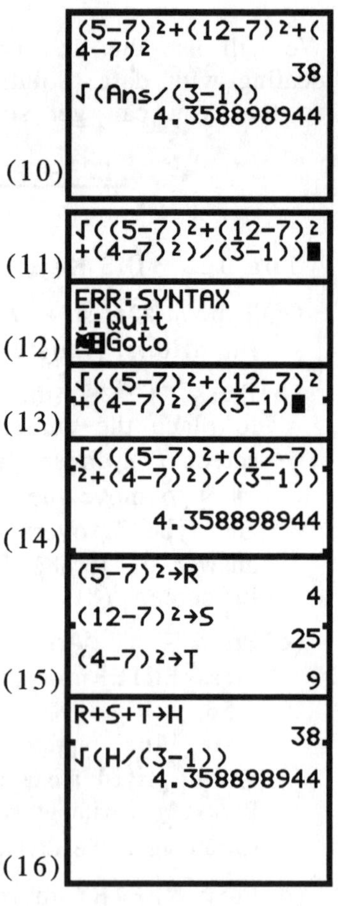


You could do the following exercises differently, but here are some ideas.

3.

```
(1.96²*(0.4)(0.6
))/0.025²
       1475.1744
1.96²*.4*.6/.025
²
       1475.1744
```

4.

```
(98.20-98.60)/(0
.62/√(106))
    -6.642342026
(98.2-98.6)/.62*
√(106)
    -6.642342026
```

5. ! under **MATH<PRB>4**.

```
25!/(16!9!)
         2042975
25*24*23*22*21*2
0*19*18*17/(9*8*
7*6*5*4*3*2*1)
         2042975
```

6.

```
√((10(513.27)-71
.5²)/(10(10-1))
     .4766783215
√((10*513.27-71.
5²)/(10*9)
     .4766783215
```

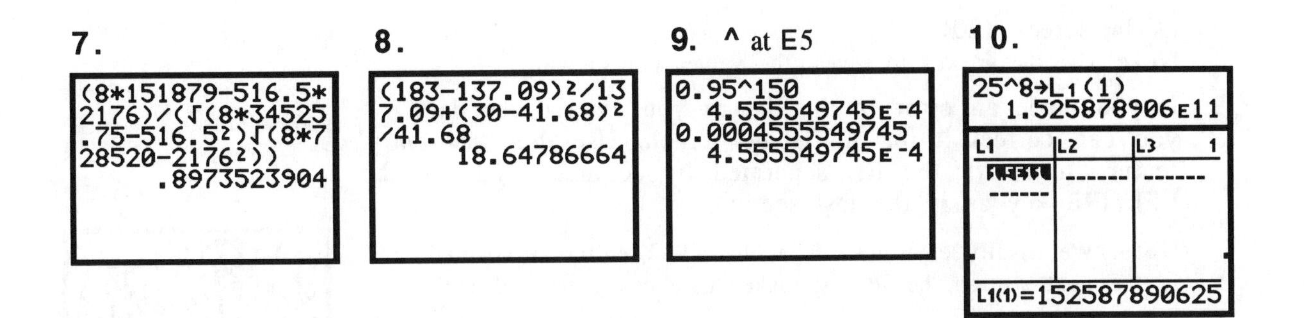


## STORING LISTS OF DATA

Most of the raw data you will use with this companion needs to be stored in lists on your TI-83 Plus. The data sets given in the Appendix B of the Triola text can be transferred to your calculator from another calculator (or from a computer) and installed in RAM with the **APPS** key. See your instructor and/or the Appendix of this companion for more information.

### Storing Lists of Data from the Home Screen

**EXERCISE 17** [pg. 24] Revisited from page 4 of this manual: Ten random integers between 1 and 28 was generated and displayed in a list on screen (5).

Modify screen (5) by repeating the first three lines but now with 10 values and storing this list as L1 by pressing **STO▶** and engage L1 above the **1** key for the forth line in screen (17). Press **ENTER** for the fifth line. Use the ▶ key to reveal the values not showing.

```
123→rand
                123
randInt(1,28,10)
→L1
{20 8 21 21 13 ...
```
(17)

To see the complete list on the home screen press **↑RCL**, above the **STO▶**key, and engage L1 for the last line of screen (17). Press **ENTER** for the last two lines of screen (18).

```
Rcl L1
randInt(1,28,10)
→L1
{20 8 21 21 13 ...
{20,8,21,21,13,1
3,24,14,19,16}
```
(18)

Use the cursor control keys and **DEL** to delete replicates and the last three values in the list. Restore in L1 for the last three lines in screen (19).

```
{20 8 21 21 13 ...
{20,8,21,13,24}→
L1
   {20 8 21 13 24}
```
(19)

**EXAMPLE** Store the five heights in inches (at the right) of students as L2 and as a list named HTS.

| 64 | 67.5 | 68 | 70.5 | 65 |
|---|---|---|---|---|

1. From the Home screen, engage the left set symbol, {, (found above the left parenthesis). Type in the data above, separating each value with a comma at B6. Press **STO▶** and engage L2 above the **2** key. Press **ENTER** for the set to be repeated without the commas,

```
{64,67.5,68,70.5
,65→L2
{64 67.5 68 70....
```
(20)

as in screen (20).
**Note**: Use the ▶ key to reveal the values not showing.

2. If you made an error in the list as you entered the data, you can correct it on the Home screen. Use the 'last entry' feature to return the list separated by commas, and use the **DEL↑INS** key as in the last section.

**Note**: We usually enter data from the STAT editor as covered in the next section, because it makes data easier to deal with.

3. To change from inches to meters, multiply each value by 0.0254 and store the results in L3, as shown in the first two lines of screen (21) with L2 **× 0.0254 STO▶** L3 **ENTER**.

(21)

**Note**: The ▶ key could be used to reveal the last values of the list or **↑RCL** as in the last exercise and in the last lines of screen (21).

4. Take values in L2 and store them in a list named HTS, as in the first two lines of screen (22), with L2 **STO▶ ↑A-LOCK HTS ENTER**.
Then type **HTS** and press **ENTER** for a single value result of 8550 instead of the list of values expected.

5. To see what happened in the last step, type **ALPHA H** and press **ENTER** for the result of 38, as in the last two lines of screen (22). Similarly, T gives 9 and S gives 25.
**Note**:These were the values saved in the last section (screens (15) and (16))—you might have different values.

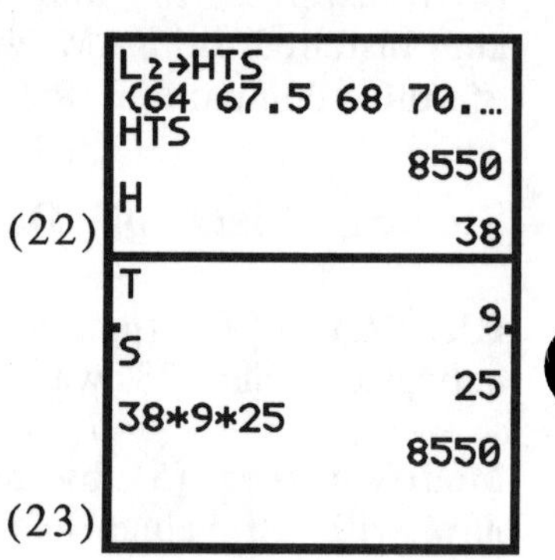

(22)
(23)

6. We need a way to distinguish between products and list names. Press **↑LIST** for a screen such as screen (24). Highlight the number, or space, before the list name HTS with the ▼ key, and then press **ENTER** for ʟHTS to be pasted to the Home screen as in screen (25).

**Note**: Under **↑LIST<NAMES>** the lists are in alphabetical order. If there are numerous list names, HTS might not be on the first screen. Pressing **ALPHA**, and then **H** advances us to the correct screen. (Pressing ▲ when at the top Name sends us to the last Name.)

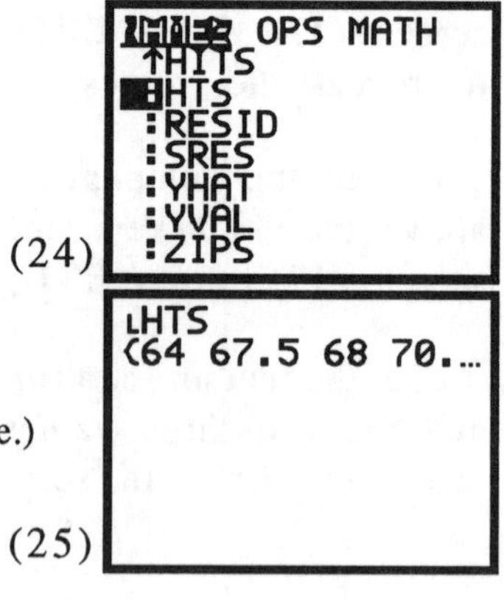

(24)
(25)

We did not need the small ʟ when we stored a list of data because it was understood that this must be stored to a list. We cannot type HTS alone on the screen, however, if we mean to designate a list. In other situations where it is not clear whether a small ʟ is needed, your safest bet is to paste from **↑LIST<NAMES>**. This listing of names also has the advantage of giving us the correct spelling of the list saved.

**Note**: The small ʟ can also be pasted to the Home screen by pressing **↑LIST<OPS> B:ʟ ENTER**, and then HTS could be typed.

## Storing Lists of Data Using the STAT Editor

**SetUpEditor:** The TI-83 Plus can use named lists in addition to L1 to L6 and is limited only by memory size. To make the Stat Editor have only L1 to L6 showing, press **STAT 5:SetUpEditor** and then **ENTER** for 'Done' as shown in the first lines of screen (26).

(26)

We want to put these data into the stat editor, but first we need to clear out old data.

1. **Clear Lists in Stat Editor.**

   Press **STAT 4**: ClrList L1,L3 **ENTER** (don't forget the comma) for 'Done' on the Home screen, indicating that L1 and L3 are cleared, as shown in the last lines of screen (26).

2. **View or Edit in the Stat Editor.**

   **STAT 1**:Edit... reveals screen (27) with L1 and L3 cleared and with the Height data in L2.

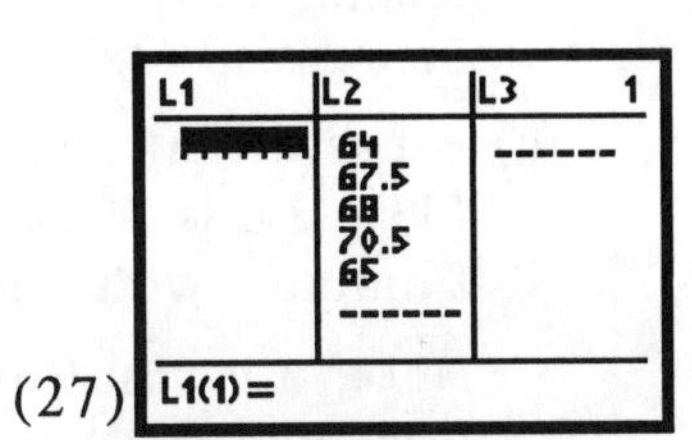

(27)

3. **Enter Data into the Stat Editor.**

   (a) With the cursor at the first row of L1, type **1 ENTER**. The cursor moves down one row.

   (b) Type **3**, and the screen will look like screen (28), with 3 in the bottom line. Press **ENTER**, and 3 will be pasted in the second row and the cursor will move down one row.

   (c) Continue with 4, 45, and 5.

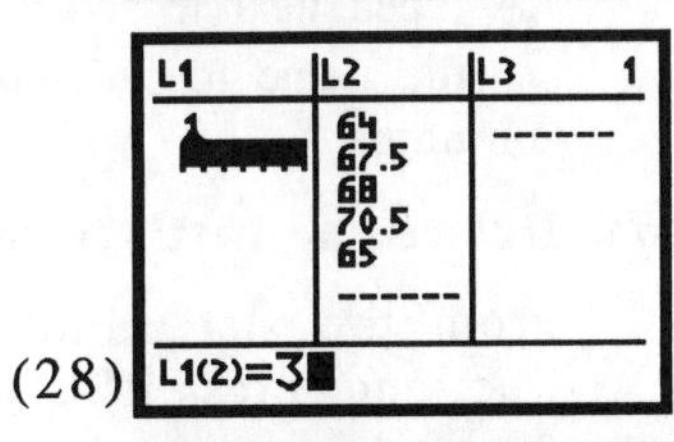

(28)

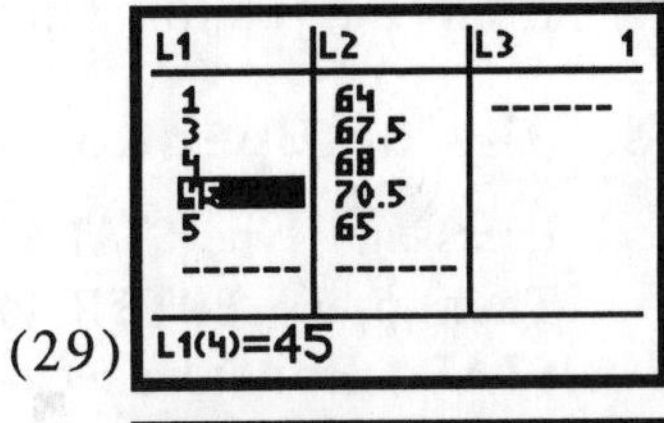

(29)

4. **Correct Mistakes with DEL and ↑INS.**

   (a) In screen (29) we can delete the 45 that we highlighted by using the ▲ key and then pressing **DEL**.

   (b) To insert a 2 above 3 move the cursor to 3 and press **↑INS** to get screen (30), with a 0 where 2 goes.

   (c) To insert a 2 type **2 ENTER**.

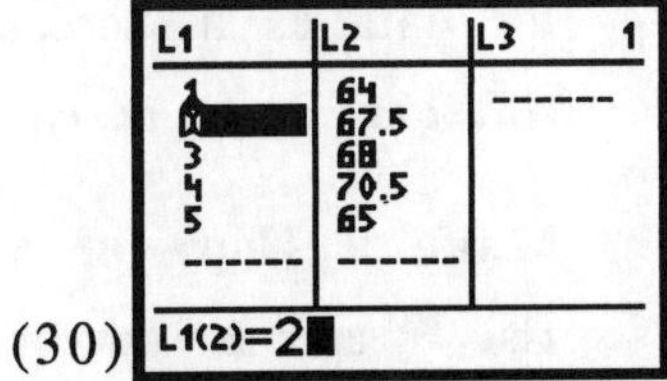

(30)

5. **Clear Stat Editor List Without Leaving It.** Use cursor control keys to highlight L2 as in the top line in screen (31). Press **CLEAR**, then ▼ (or **ENTER**). Pressing **CLEAR** reduces the last line to L2= as in screen (31), but the data still exist in L2 until you press ▼ (or **ENTER**); then they are gone (not shown).

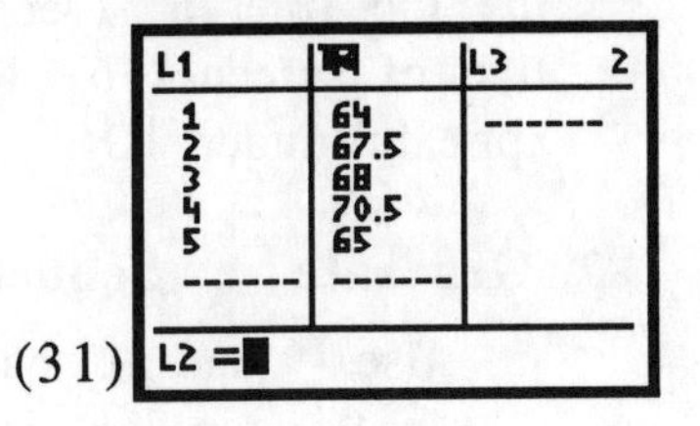

(31)

**Note:** I often forget to clear a list before entering the stat editor as in step **(1)**, so I tend to clear the list this way. Both methods require about the same number of keystrokes, however.

**6. Store Data with a Named List.**

Let's store the random numbers 43, 26, 58, 40, 14, 71, and 20 from the Lotto example on page 5 in a new list called RAND2.

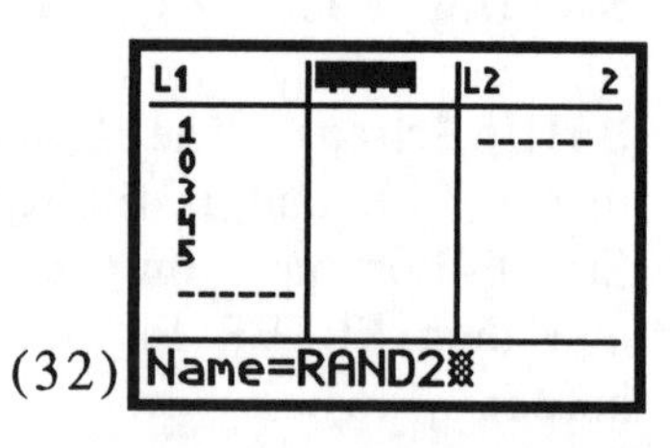

(32)

(a) First with L2 highlighted, press ↑INS, which moves list L2 to the right and prompts for a Name in the bottom row in ALPHA mode, as in screen (32).

(b) Type RAND and then press **ALPHA** (to deactivate) and **2**. A flashing quilt pattern will appear indicating that we have used the maximum name length of five characters, as shown in the last line of screen (32).

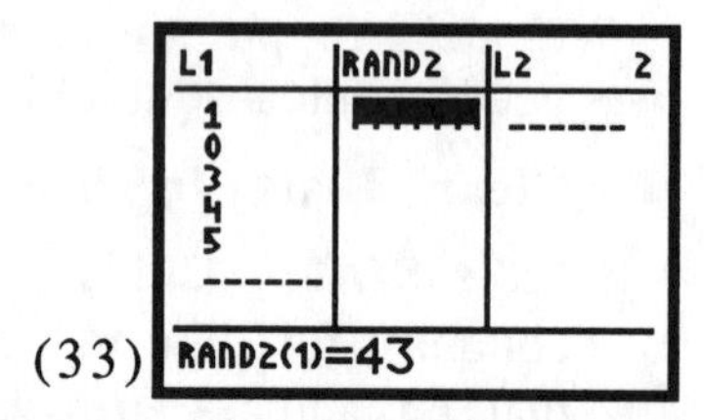

(33)

(c) Press **ENTER** and then ▼ and type **43** for screen (33). Press **ENTER**, and 43 is pasted in the first row. Continue with the other six values.

**Note:** If the list RAND2 had already been created, its name could have been pasted or typed next to 'Name' in screen (32). Pressing **ENTER** then would not only have pasted the name to the top line but also pasted the data (if it existed) in the rows below the name.

**7. Delete a List from the Stat Editor.**

From the stat editor, highlight the name on the top line and press **DEL**. The name and the data are gone from the editor but not from memory.

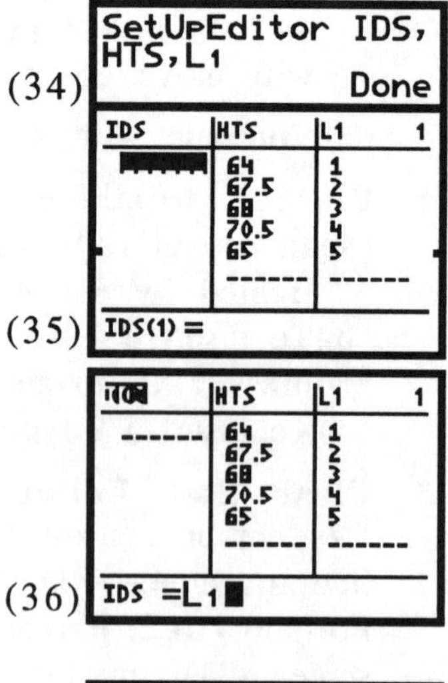

(34)

(35)

(36)

**8. Use SetUpEditor to Name List.**

Press and type **STAT 5**: SetUpEditor IDS, HTS,L1. Then press **ENTER** for 'Done' as in screen (34). Press **STAT 1**:Edit to reveal the new list named IDS ready for data as in screen (35).

**Note:** HTS can be typed or pasted from **↑LIST<NAMES>**.

**9. Make a Copy of a List.**

Use ▶ and ▲ cursor control keys to highlight IDS as in the top line in screen (36). Engage L1 as in the bottom line of screen (36). Press **ENTER** and the L1 data appears under IDS.

**10. Generate a Sequence of Numbers In a List.**

Use ▶ and ▲ cursor control keys to highlight L1 as in the top line in screen (37). Press **↑LIST<OPS> 5:seq(X,X,1,28** as in the bottom line of screen (38). Press **ENTER** for the sequence of integers from 1 to 28 as in screen (38).

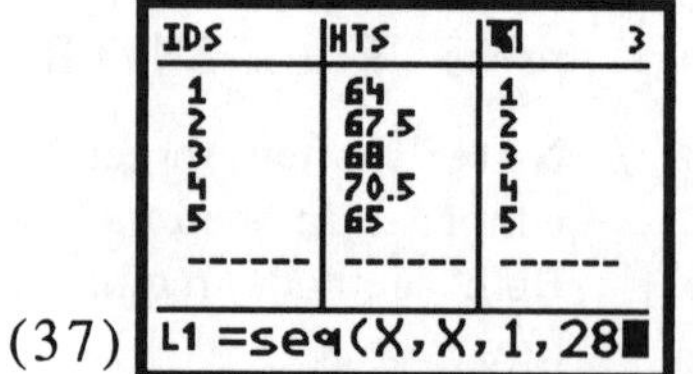

(37)

**Note:** To check the values of data in a spreadsheet, you can jump a page at a time down or up the list with each press of the green **ALPHA** key followed by either ▼ or ▲ . (The green arrows between these cursor control keys are a visual reminder of this capability.)

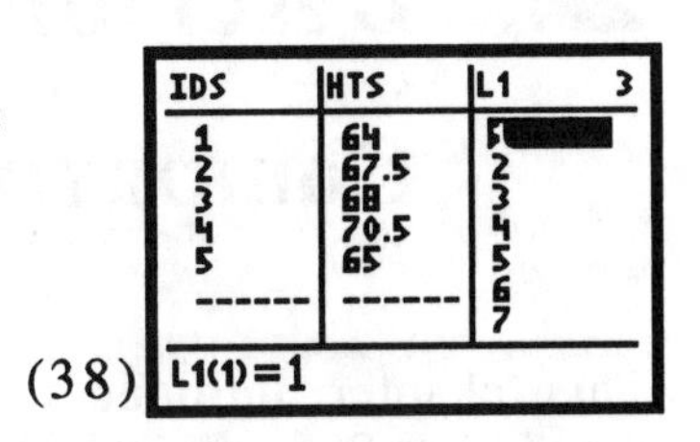

(38)

**Note:** Random integers, like those generated on page 5, could also be generated in the Stat Editor similar to the above.

## 11. Delete a Named List from Memory.

(a) To remove both the name and data from RAM (random access memory), press ↑**MEM** (above the + key) for screen (39).

(b) Press **2**:Mem Mgmt/Del (or **2**:Delete... on the TI-83) for a screen like (40).

(c) Press **4**:List... for a display of list names. Use ▼ to move the selection cursor to the list you want to remove (say IDS), as shown in screen (41). Be careful!

(d) Press **DEL** (or **ENTER** on the TI-83) to delete the list. You can remove lists one by one from this screen.

(e) Press ↑**QUIT** to return to the Home screen.

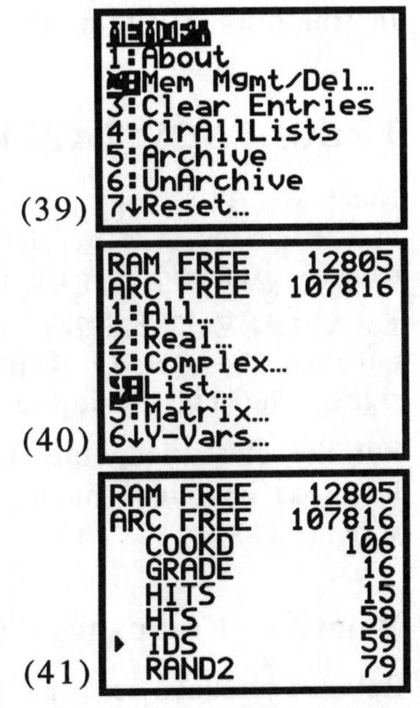

(39)

(40)

(41)

**Note:** (For TI-83 Plus only) If the list was also saved in a Group stored in ARCieve memory, the above procedure has not deleted that copy. (See the Appendix, page 81, for more information on Groups.) Groups can be deleted from the calculator if in step (c) above you replace **4**:List... with **8**:Group... .

## 12. For Large List, ↑LIST<OPS>9:augment Is Useful.

Several people can cooperate in keying in large data sets. Say part is stored in L1 on one TI-83 Plus and another part is stored in L2 on another TI-83 Plus. These data can be shared by linking the two machines and then combining L1 and L2 and storing the results in L3 with ↑**LIST<OPS> 9**:augment(L1,L2)**STO▸**L3. This procedure can be extended to more than two people.

When storing small data sets for temporary use, using L1 to L6 is convenient. However, it would be wise to name large data sets because you have invested more time to store them.

# 2 Describing, Exploring, and Comparing Data

This chapter introduces the plotting and summary statistics capabilities of the TI-83 Plus. ↑**STAT PLOTS** and other keys in the first row are for descriptive plots. Pressing **STAT** <CALC> **1:** 1-Var Stats gives summary statistical calculations. Be sure you have studied Chapter 1 so that you can follow the more abbreviated instructions in this chapter.

## FREQUENCY TABLES [pg. 35]

**Note:** The TI-83 Plus can automate the construction of a frequency table by plotting a histogram from raw data, as explained on page 15.

**EXAMPLE** [Table 2-3, pg. 35]: The frequency table of the Qwerty word ratings is repeated at right. The class midpoints are included.

| Class limits<br>**Rating** | L1<br>**Midpoints** | L2<br>**Freq** |
|---|---|---|
| 0-2 | 1 | 20 |
| 3-5 | 4 | 14 |
| 6-8 | 7 | 15 |
| 9-11 | 10 | 2 |
| 12-14 | 13 | 1 |

Put the class midpoints in L1 and the frequencies in L2, as explained on page 9, and shown in screen (1).

### Relative Frequency Table [Table 2-5, pg. 38]

Relative frequencies can be calculated in L3.

1. Highlight L3 at the top of the column, as in screen (1).
2. Type L2÷**52×100** for the bottom line of screen (1).
3. Press **ENTER** for screen (2). Notice that the first cell (centered at 1) has about 38.5% of the data, or 20/52*100 = 38.46, as shown in the first row of L3 (but easier to read from the bottom line of the screen (2)).

   **Note:** You could have used L2⁄sum(L2 instead of L2⁄52 in step **2**, with sum pasted from ↑**LIST**<MATH>.

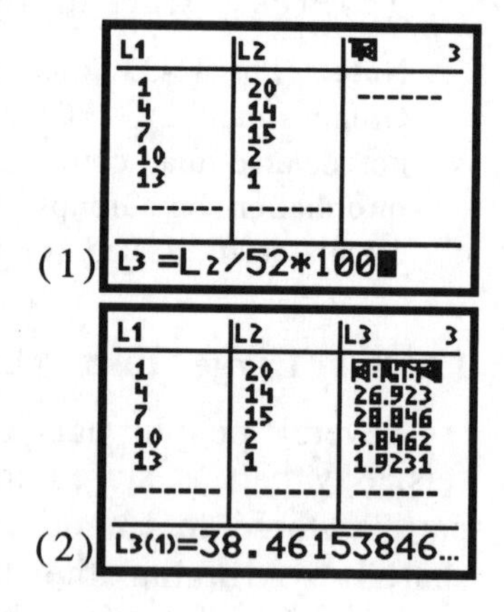


### Cumulative Frequency Table [Table 2-6, pg. 39]

Cumulative frequencies can be calculated in L4.

1. Use the ▶ key to continue beyond L3 to L4. Highlight L4 at the top of the column, as in screen (3).
2. Press ↑**LIST**<OPS>**6:**cumSum(L2 for the bottom line of screen (3).

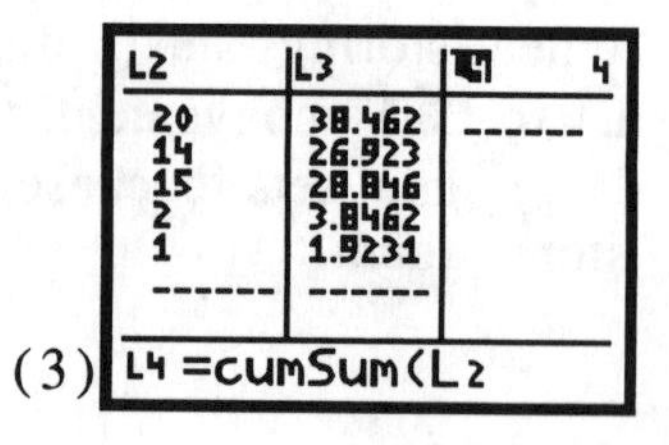

3. Press **ENTER** for screen (4). Notice that the second row indicates that 20 + 14 = 34 values are in the first two cells, and so on.

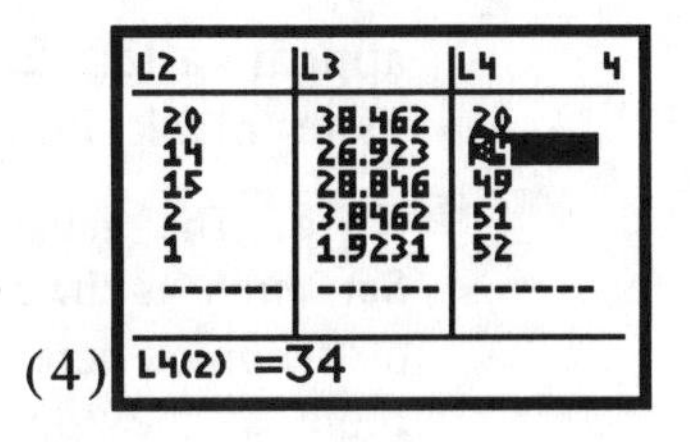

(4)

**EXERCISES 13–20** [pg. 41]: Construct a frequency table using the raw data given in the text.

The TI-83 Plus can automate the construction of a frequency table by plotting a histogram from raw data, as explained on page 15.

## HISTOGRAMS FROM FREQUENCY TABLES [pg. 42]

**FIGURE 2-2** [pg. 43]: This example continues with the frequency table of qwerty word ratings of the last section with class marks in L1 and frequencies in L2.

1. **Turning OFF All Stat Plots.**
   Activate **↑STAT PLOT**, in the upper left corner of the keyboard, for a screen such as screen (5). If all plots but Plot1 are not 'Off', press **4**:PlotsOff **ENTER** for 'Done.'

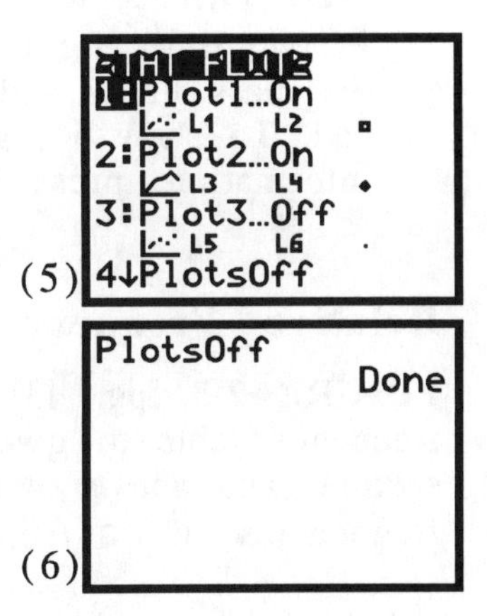

(5) (6)

**Note**: You also need to make sure that all plots on the **Y=** edit screen, if you have been using this, are off or cleared. To clear an equation, move the cursor to the right of the equal sign for that function and press **CLEAR**.

2. **Turning ON and Setting Up Stat Plot1.**
   (a) Engage **↑STAT PLOT 1**: Plot1... for a defining screen which will be set up as on screen (7).
   (b) Use the ▼ ▶ cursor control keys to highlight the choices on screen (7), and press **ENTER** after each choice to activate it. Note that "Type" is a histogram, the third choice in the first row of types, "Xlist" shows the class marks in L1, and the "Frequencies" are in L2.

(7)

3. **Setting Up Plot Windows.**
   (a) Press the **WINDOW** key (at B1) for a screen with all but the numbers, as screen (8).
   (b) At the Xmin= line, type **⁻0.5** for the lowest class boundary from the frequency table.
   (c) Press **ENTER** to advance to Xmax=. Enter the highest class boundary, **14.5**,
   (d) Let Xscl=3 since 3 is the width of the cells (e.g., 4 – 1 = 3, the distance between class midpoints).
   (e) Since the maximum frequency is 20, make Ymax=24 and thus Ymin=⁻24/4, or ⁻6, which is the way this will

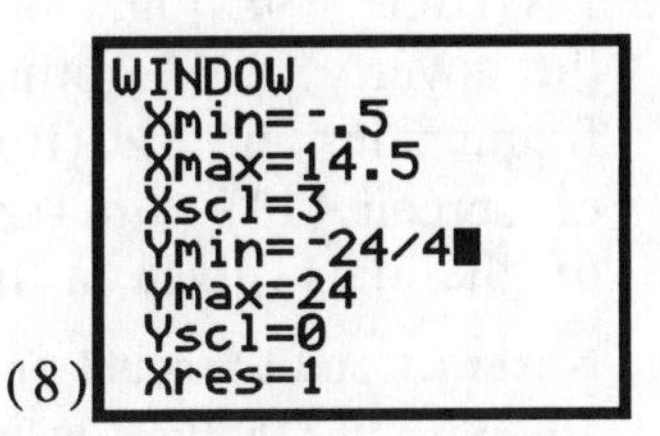

(8)

appear after **ENTER** is pressed in the display screen (8). (The black box cursor is waiting for **ENTER**.)

**Note:** The negative symbol is in the bottom row of the keyboard. Set Ymin as the negative of Ymax ÷ 4 to leave room at the bottom of the plot screen for the cell information. Also leave extra room at the top for plot setup information.

(f) Yscl sets the scale of the Y-axis. It was set at 0, so no tick mark will show on the Y, or frequency, axis.

(g) Let Xres=1 or all WINDOWs in this companion.

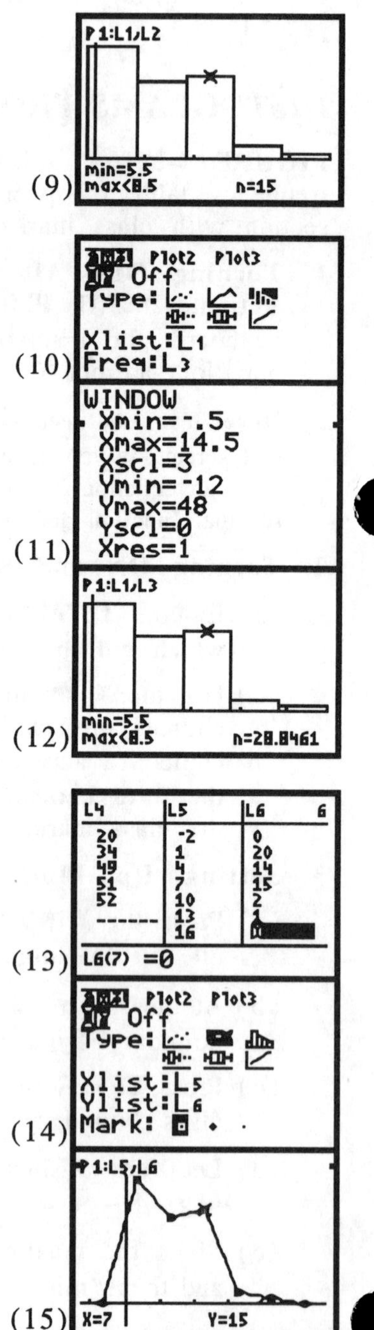


**4. Ploting the Histogram.**

Press **TRACE**, wait for the plot to appear, and then press ▶ two times for the histogram in screen (9). Notice the class limits and the frequency, 15, for the selected cell. To show a graph screen without the added information, press **GRAPH** or **CLEAR**.

## Relative Frequency Histogram [pg. 43]

**FIGURE 2-3** [pg. 43]: This example continues with the frequency table of qwerty word ratings of the last section (and above) with class marks in L1 and relative frequencies in L3 (screen (2).

Adjusting the plot and window setups as shown in screens (10) and (11) gives the results of screen (12), after pressing **TRACE**, wait for the plot to appear, and then press ▶ two times for n = 28.8% of the words having ratings from 6 to 8.

## Frequency polygons [pg. 43]

**FIGURE 2-4** [pg. 44]: Construct a frequency polygon for the qwerty word ratings with class midpoints in L1 and frequencies in L2 (from screen (1)) repeated in L5 and L6 of screen (13) with an extra class midpoints on each side of the data given a frequency of zero.

**Note:** L1 and L2 could be copied to L5 and L6, (as explained on page 10), then the bottom values of 16 and 0 added. Next insert the top values of -2 and 0 (as explained on page 9).

With the data in L5 and L6, set up Plot1 for an xyLine plot as in screen (14). Press **ZOOM** 9:ZoomStat and then **TRACE** for the frequency polygon plot in screen (15).

## Ogive [pg. 43]

**FIGURE 2-5** [pg. 44]: Construct an ogive for the qwerty word ratings with the cumulative frequencies in L4 (from screen (4)) repeated in L6 of screen (16) with an initial 0 inserted. The class boundaries in L5 begins with the lower boundary of the first class of -0.5, and ends with the upper boundary of the last class of 14.5.

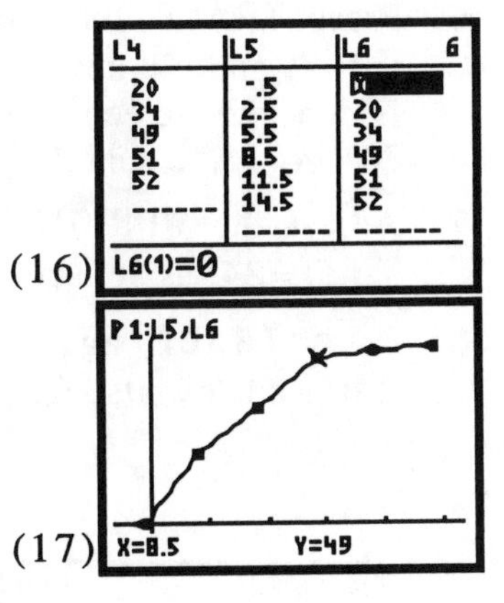

(16)
(17)

With the data in L5 and L6, set up Plot1 for an xyLine plot as in screen (14) above. Press **ZOOM 9**:ZoomStat and then **TRACE** for the ogive plot in screen (17). The bottom line of screen (17) shows that there are 49 values less than 8.5.

## HISTOGRAMS AND FREQUENCY TABLES FROM RAW DATA

**TABLE 2-1** [pg. 33]: The querty word ratings are repeated at right.

Qwerty Keyboard Word Ratings

| 2 | 2 | 5 | 1 | 2 | 6 | 3 | 3 | 4 | 2 |
|---|---|---|---|---|---|---|---|---|---|
| 4 | 0 | 5 | 7 | 7 | 5 | 6 | 6 | 8 | 10 |
| 7 | 2 | 2 | 10 | 5 | 8 | 2 | 5 | 4 | 2 |
| 6 | 2 | 6 | 1 | 7 | 2 | 7 | 2 | 3 | 8 |
| 1 | 5 | 2 | 5 | 2 | 14 | 2 | 2 | 6 | 3 |
| 1 | 7 | | | | | | | | |

1. Enter the data into list LQWERT, as explained on page 10, with numbers entered across by row-order. Or the data could be transferred from another calculator or from a Group (see Appendix page 81).

2. Under ↑**STAT PLOT** after selecting and turning Plot1 On with "Type" being the histogram as in screen (19), set
   (a) "Xlist" at QWERT, and
   (b) "Freq" at **1**. This setting means we are going to count each value in QWERT, one time. There are two 10s in QWERT, and each one is counted once and put in the proper cell; in other words, the frequency of that cell is increased by 2.

3. **Automatic Histogram**
   Press **ZOOM 9**:ZoomStat and then **TRACE** for screen (20) with 5 words rated 0 or 1. By using the ▶ key, you can find the frequencies in the other seven cells.

4. This is a good first look at the data, but you might want to set your own cell limits. Press **WINDOW** for screen (21). (In other examples you will see that cell widths are usually not integers such as the Xscl = 2 of screen (21).)

5. Change the values in the **WINDOW**, and set the cell width, or Xscl = 3, as in screen (22). Set Xmin to the lower boundary of the first class or -0.5, and Xmax with the upper boundary of the last class or 14.5.

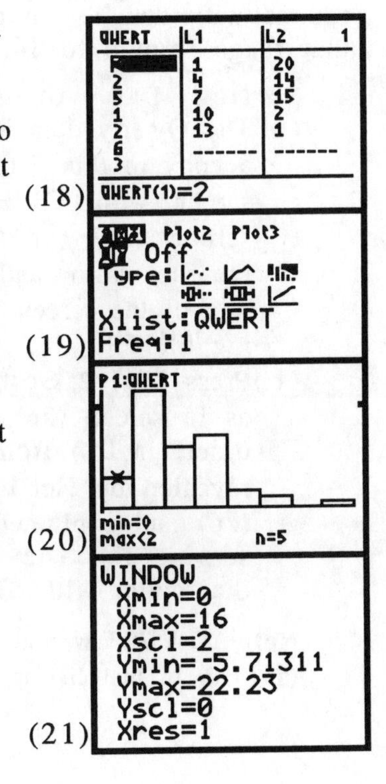

(18)
(19)
(20)
(21)

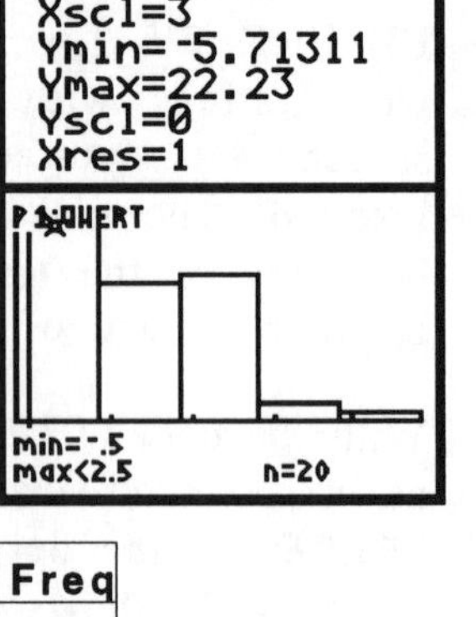


6. Press **TRACE** for the histogram in screen (23). By using the ▶ and ◀ keys, even though the histogram does not fit on the screen, we find the five cells have values of 20, 14, 15, 2, and 1.
7. Adjust the **WINDOW** height by letting Ymin= ⁻6 and Ymax=24, as in screen (24) below.
8. Press **TRACE** again for the histogram in screen (25) followed by its frequency table.

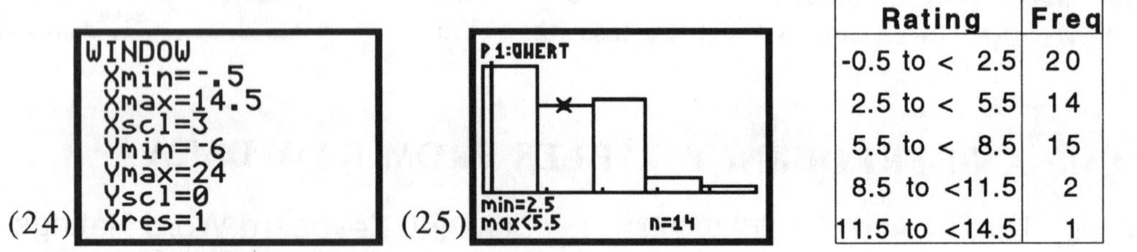


| Rating | Freq |
|---|---|
| -0.5 to < 2.5 | 20 |
| 2.5 to < 5.5 | 14 |
| 5.5 to < 8.5 | 15 |
| 8.5 to <11.5 | 2 |
| 11.5 to <14.5 | 1 |

## DOTPLOTS AND STEM-AND-LEAF PLOTS [pg. 44]

A dotplots and Stem-and-leaf plots are not Plot Types on the TI-83 Plus but they are not difficult to do by hand if the data is put in order with the TI-83 Plus and the data set is not too large.

**FIGURE 2-6** [pg. 44]: Construct a dotplot of the qwerty word ratings with data as on page 15. The data values are integers that range from 0 to 14.

1. **Sorting Data in Ascending Order.**
   (a) The Qwerty data is currently listed in ʟQWERT. Make a copy of this data by storing it in L1 from the Home screen, with ʟQWERT **STO▶** L1, as in the top lines of screen (26) (or alternately by highlighting L1 in the Stat editor and pasting ʟQWERT in the bottom line of the screen for L1 = ʟQWERT and then pressing **ENTER**.
   (b) Press **STAT** 2: SortA(L1 and then **ENTER** for 'Done' as in screen (26). The data is now in ascending order (in L1) from low value to high. This is revealed by Rcl L1 as in the bottom line of screen (26) and explained on page 7 with the results as in screen (27). The data could also be observed in the Stat editor with **STAT** 1:Editor. See screen (28).

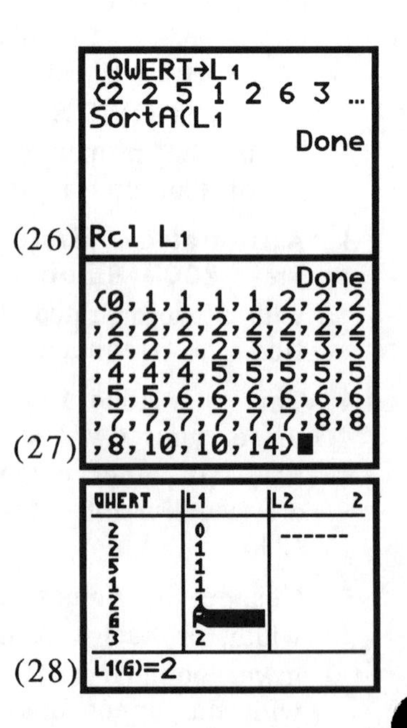


**Note:** ʟQWERT was not sorted directly because the order of the data in that list is important for future work.

2. Count the frequency that each data value occurs. There is one word rated 0, four words rated 1 and by continuing down the list you can count the number or words with each rating. '2' is the most frequent with 15 values and there are no words rated 9, 11, 12, or 13 with only two rated 10 and one rated 14. These counts are nicely shown with dots in the plot at the right although dotplots are usually shown with the dots going up from the number line.
   Since the data is mostly one digit integers the Stem-and Leaf Plot would look basically like the plot at the right but with the dots replaced by leafs of all zeros or '0'. An example with more variety follows.

```
0  |*
1  |****
2  |***************
3  |****
4  |***
5  |*******
6  |******
7  |******
8  |***
9  |
10 |**
11 |
12 |
13 |
14 |*
```

**Note:** Since the data above are all integers between zero and 14 a histogram with cell width one (Xscl = 1) and with other window values of Xmin = -.5 and Xmax = 14.5 shows nicely the frequency of each word rating value. Try it!

**EXAMPLE** [pg.45]: Construct a stem-and leaf plot for the following test grades.

67 72 85 75 89 89 88 90 99 100

```
{67,72,85,75,89,
89,88,90,99,100→
L1
{67 72 85 75 89…
SortA(L1
                Done
Rcl L1
```
(29)

This data set is saved and sorted in L1 (a copy was not made because the data set was small and the original order was not important). See screen (29). The results were recalled to the home screen as in the last example for the results in screen (30).

```
89,88,90,99,100→
L1
{67 72 85 75 89…
SortA(L1
                Done
{67,72,75,85,88,
89,89,90,99,100}
```
(30)

From screen (30) it is easy to read the scores in order with the first score of 67 having a stem of 6 and a leaf of 7. There are two grades of seventy so a stem of 7 with two leafs of 2 and 5. Continuing in this manner to the highest grade of 100 with a stem of 10 and a leaf of 0 as in the plot at the right and in the text.

```
 6|7
 7|25
 8|5899
 9|09
10|0
```

## PARETO AND BAR CHARTS [pg. 46]

**FIGURE 2-7** [pg. 46]: In a recent year, 75,200 accidental deaths in the United States were attributable to the accident types in the table at the right.

| Accident Type | Deaths |
|---|---|
| Motor vehicles | 43500 |
| Falls | 12200 |
| Poison | 6400 |
| Drowning | 4600 |
| Fire | 4200 |
| Ingestion | 2900 |
| Firearms | 1400 |

1. Put the seven values 1, 2, 3, 4, 5, 6, 7, in L1.
2. Put the number of deaths for the seven types into L2 as shown in screen (31).

   **Note**: If the data were not in descending order from

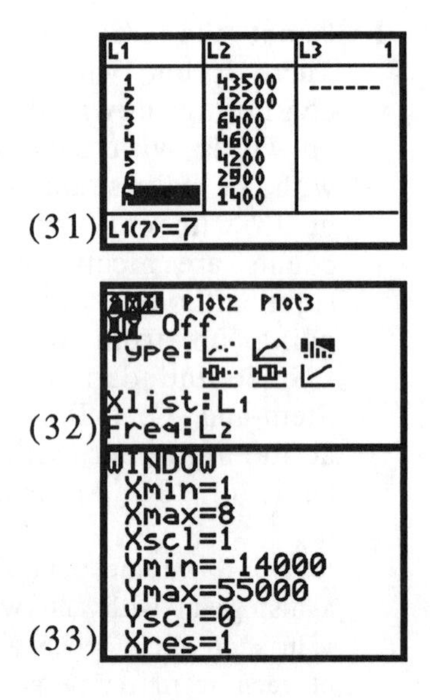


high to low value, use **STAT 3:SortD(L2** (similar to what was done in the previous example).

3. Set up Plot1 as a histogram in screen (32), and the **WINDOW** as in screen (33), and press **TRACE** for the Pareto chart in screen (34) below.

4. A bar chart of the above data has a gap between the different categories. This gap is obtained by changing Xscl = 0.5 in the WINDOW of screen (33) and then pressing **TRACE** for screen (35).

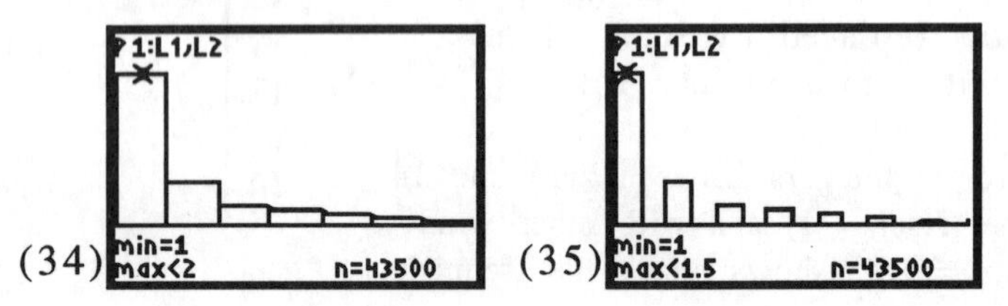


## PIE CHARTS [pg. 46]

**FIGURE 2-8** [pg. 46]: The TI-83 Plus **STAT 1**:Editor can help us make a pie chart of the 75,200 accidental deaths from the example above by calculating the percentages for each type and the degrees in the central angles.

1. With the data in L2, highlight L3, as in screen (36). Type **100L2÷sum(L2** as in the bottom line of the same screen, with 'sum(' from **↑LIST<MATH>5:sum(**.

2. Press **ENTER** for the percentages of each type of accident in L3, as in screen (37). Highlight L4, in the top line, and type **360L2÷sum(L2** as in the bottom line of the same screen.

3. Press **ENTER** for the center angle in degrees of each type of accident in L4, as shown in screen (38).

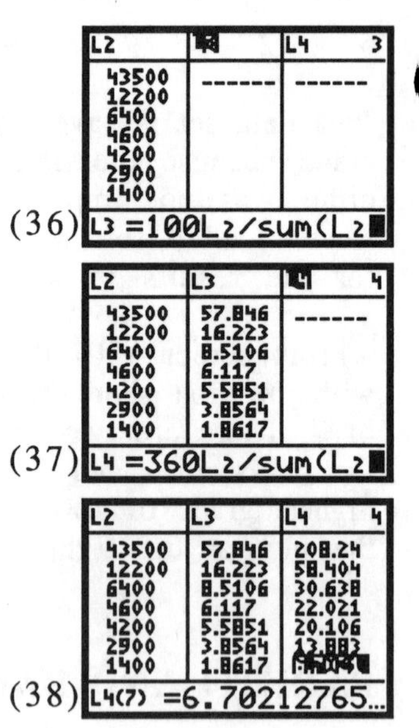


To summarize, in the first row of screen (38), the 43500 deaths due to motor vehicles represent 57.8% of the deaths, or 208 degrees of the pie.

## SCATTER DIAGRAMS [pg. 47]

The first plot in chapter nine is a scatter diagram so the construction will be covered there.

## FREQUENCY POLYGONS [pg. 50]

**FIGURE 2-10** [pg. 50]: Compare the keyboard word ratings for both the Qwerty and Dvorak keyboards by constructing two frequency polygons.

Construct a frequency polygon for the qwerty word ratings by repeating lists L5 and L6, of screen (13), and the plot setup of screen (14). Construct a frequency polygon for the dvorak word ratings by putting the mid-points in L1 and the frequencies in L2 as in screen (39). (Note all values lie in just two cells.)

Set up Plot2 as in screen (40), leaving Plot1 On, and use a different symbol then the square of Plot1.
Press **ZOOM** **9**:ZoomStat and then **TRACE** for the frequency polygon plots in screen (41). Use ▲ to move between plots.

**Note:** Up to three plots can be compared on one screen.

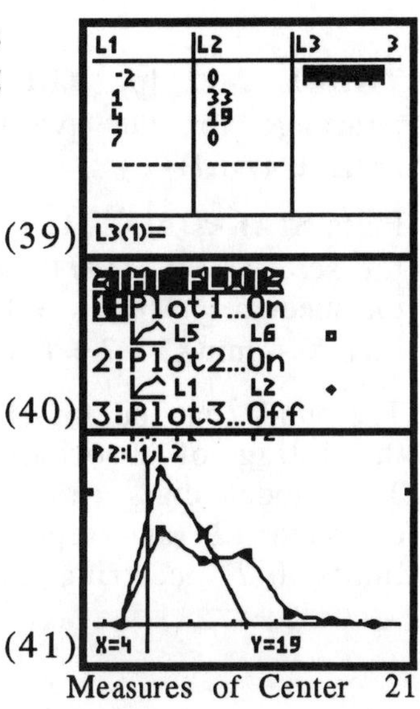


(39)

(40)

(41)

Measures of Center 21

## MEASURES OF CENTER [pg. 55]

**EXAMPLE Executive Women** [pg. 57]: Listed below are the salaries (in millions of dollars) paid to female executives. Find the mean and median for this sample.

6.72 3.46 3.60 6.44

### Mean and Median from Raw Data

1. Put the data in L1.
   Press **STAT** <CALC> **1**: 1-Var Stats **L1** for screen (43).

   **Note**: Pressing **STAT** shows the EDIT menu, ▶ shows the CALC menu, as in screen (42), and **1** pastes "1-Var Stats" to the Home screen. Then engage **L1** for screen (43).

2. Press **ENTER** for the first of the two screens of output, screen (44). (We will call these the first and second screens of output.) To reveal the second screen of output, screen (45), hold down the ▼ cursor control key.

The mean = $\bar{x} = \Sigma x / n$ = 20.22/4 = 5.055 on the first screen of output, and the median = Med = 5.02 on the second screen of output. We will discuss more of the 1-Var Stats output later.

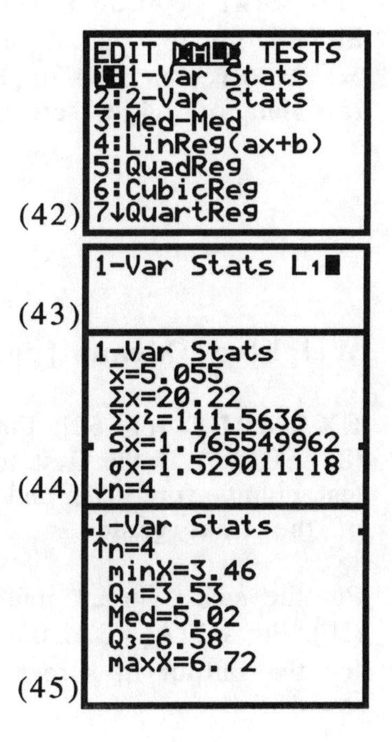


(42)

(43)

(44)

(45)

## Midrange and Mode [pgs. 58 and 59]

**TABLE 2-11** [pg. 60]: Find the mean, median, mode, and midrange for the qwerty keyboard word ratings saved in list LQWERT.

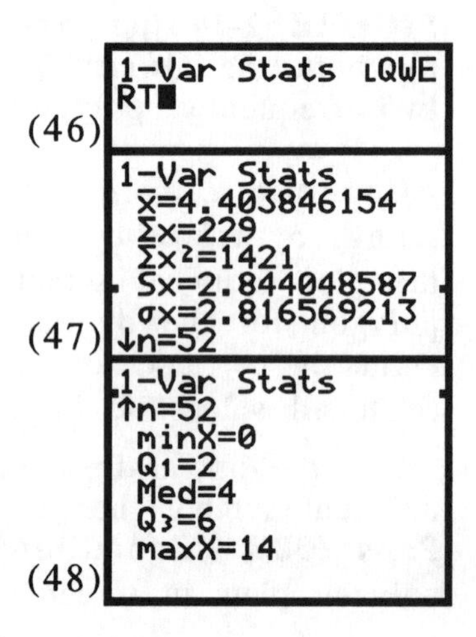

(46) (47) (48)

Press **STAT** <CALC> **1**: 1-Var Stats LQWERT and then **ENTER** for screens (46), (47), and (48) with the mean = $\bar{x}$ = 4.4, the median = Med = 4.0, and the midrange = 7.0 = (maxX + minX) ÷ 2 = (14 + 0) ÷ 2 = 7.0.

The mode can be found with the dot plot on page 17 with the rating of 2 occurring most frequently—fifteen times. The model class can be found from the frequency table on page 12 (and repeated below) with the first class with limits 0-2 occurring most frequent—twenty times.

## Mean from a Frequency Table [pg. 61]

**TABLE 2-12** [pg. 61]: The frequency table of the qwerty word ratings is repeated from page 12 in the table at the right with class midpoints stored in L1 and frequencies in L2.

| Class limits<br>**Rating** | L1<br>**Midpoints** | L2<br>**Freq.** |
|---|---|---|
| 0-2 | 1 | 20 |
| 3-5 | 4 | 14 |
| 6-8 | 7 | 15 |
| 9-11 | 10 | 2 |
| 12-14 | 13 | 1 |

```
1-Var Stats L1,L2
```

```
1-Var Stats
 x̄=4.115384615
 Σx=214
 Σx²=1348
 Sx=3.027027573
 σx=2.997780244
↓n=52
```
(49)

Press **STAT** <CALC> **1**: 1-Var Stats **L1,L2** and then **ENTER** for the output in screen (49), with mean = $\bar{x}$ = 4.1 compared to the exact value from the raw data of 4.4 in screen (47).

## Weighted Mean [pg. 62]

**EXAMPLE** [pg. 62]: Find the mean of three test scores (85, 90, 75) if the first tests count for 20%, the second test counts for 30%, and the third test counts for 50% of the final grade

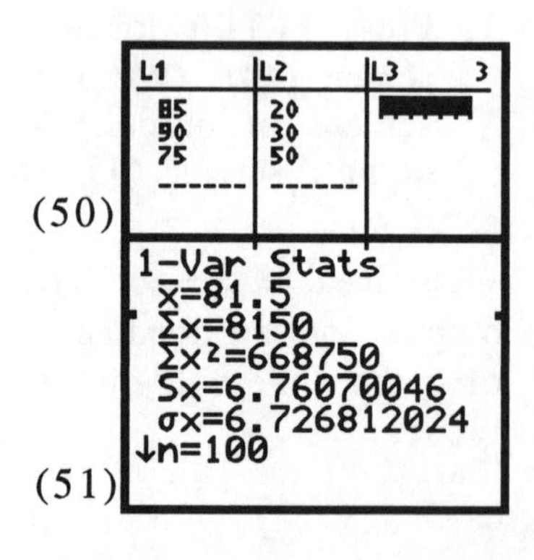

(50) (51)

Put the scores in L1 and the weights in L2, as in screen (50). Press **STAT** <CALC> **1**: 1-Var Stats **L1,L2** and then **ENTER** for the output in screen (51) with the weighted mean = $\bar{x}$ = 81.5

# MEASURES OF VARIATION [pg. 68]

**EXAMPLE** [pgs. 69, 71, and 74]: The waiting times (in minutes) for a sample of ten customers at the Jefferson Valley Bank are
6.5 6.6 6.7 6.8 7.1 7.3 7.4 7.7 7.7 7.7.

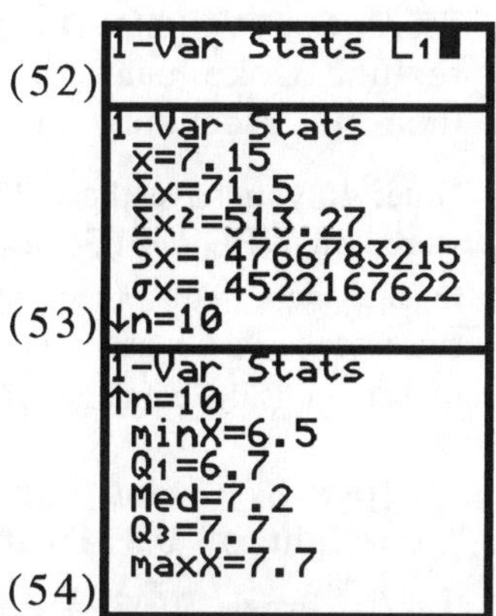


**1. Standard Deviation and Variance.**
With the data in L1, press
**STAT** [CALC] **1**:1-Var Stats L1 **ENTER**. This sequence gives screens (53) and (54), with the standard deviation of Sx=0.4766783215 = 0.48 min.
Variance = $Sx^2$ = $0.4766783215^2$ = 0.227222 = 0.23 $min^2$.

**2. Range** = maxX - minX = 7.7 - 6.5 = 1.2 minutes, which is easy enough to see with the data in order or a small data set but difficult with unordered data and larger data sets.

**3. VARS 5:Statistics Menu.**

(a) Press **VARS** at D4 for the VARS menu of screen (55).
(b) Press **5** for the Statistics sub menus of screen (56).
(c) Press **3**, and Sx is pasted to the Home screen.

**Note**: In short form; **VARS 5**:Statistics **3**:Sx.

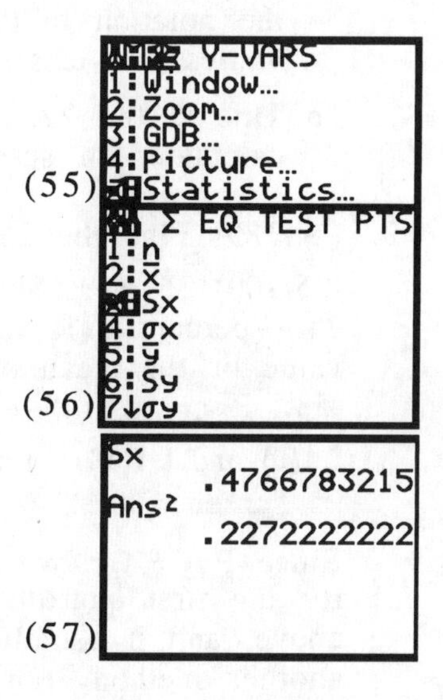


(d) Press **ENTER** for 0.4766783215, as in screen (57) and Step 1 above.
(e) Press the $\mathbf{x^2}$ key at A6 for Ans² (or Sx²) and then **ENTER** for .2272222222 as before.

**Note**: This saves typing in all the digits of Sx or rounding it off, but pasting Sx makes sense only after you have performed 1-Var Stats or another appropriate calculation; otherwise, some past calculation from other data will be stored in Sx.

## Standard Deviation from a Frequency Table [pg. 75]

**EXAMPLE: Qwerty Keyboard Ratings** [pg. 75] Find the standard deviation of the 52 values summarized in the frequency table on page 12.

This is a repeat of the procedure for finding the mean on page 20 with its output in screen (49) — also repeated here. With class midpoints stored in L1 and frequencies in L2 use **STAT** <CALC> **1**: 1-Var Stats **L1,L2** and then **ENTER** for the output of screen (49). (Sx = 3.0 compared to the exact value from the raw data of 2.8 in screen (47).)

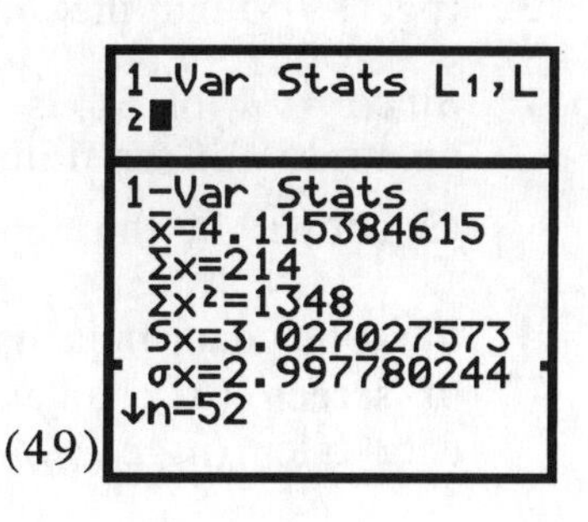

# MEASURES OF POSITION [pg. 85]

## Quartiles, Deciles, and Percentiles [pg. 86]

**EXAMPLE and TABLE 2-16** [pg. 87]: Store the weights (in pounds) of 36 regular Coke cans in row order or sort them in ascending order in list L1.

Note: If you have these [from Appendix B] stored in ʟCRGWT (Coke regular weight) or in the Data Apps (see Appendix page 80), store them in L1 and then SortA(L1) similar to screen (26) on page 16.

| | | | | | |
|---|---|---|---|---|---|
| 0.7901 | 0.8044 | 0.8062 | 0.8073 | 0.8079 | 0.811 |
| 0.8126 | 0.8128 | 0.8143 | 0.815 | 0.815 | 0.8152 |
| 0.8152 | 0.8161 | 0.8161 | 0.8163 | 0.8165 | 0.817 |
| 0.8172 | 0.8176 | 0.8181 | 0.8189 | 0.8192 | 0.8192 |
| 0.8194 | 0.8194 | 0.8207 | 0.8211 | 0.8229 | 0.8244 |
| 0.8244 | 0.8247 | 0.8251 | 0.8264 | 0.8284 | 0.8295 |

1. [pg. 87]: Find the percentile corresponding to the weight of 0.8143 lb.

   (a) Press **STAT 1**:Edit and use the ▼ cursor control key to go to .8143. This is the 9th value, as shown in the notation in the top line of screen (58), so 8 values are less.

   (b) This is the 22nd percentile of can weights as calculated in screen (59) with (8/36)*100 = 22.

```
L1      L2      L3      1
.8143
.815
.815
.8152
.8152
.8161
.8161
L1(9)=.8143
```
(58)

```
(8/36)*100
          22.22222222
```
(59)

2. [pg. 89] Find the 25th percentile, or $P_{25}$.

   (25/100)*36 = 9. Since this is a whole number, the 25th percentile is midway between the 9th and 10th value or the mean of the top two values in screen (58). $P_{25}$.= (0.8143 + 0.8150)/2 = 0.81465 or we can use **L1(9)** and **L1(10)** for the 9th and 10th value as shown in screen (60).

   Since $P_{25} = Q_1$, we would hope that the TI-83 Plus value for the first quartile is close to what we calculated above, and it usually is, although we calculated it by another method. For this example, they agree exactly as the second output screen, (61), from **STAT** <CALC> **1**: 1-Var Stats.

```
(25/100)*36
                    9
(L1(9)+L1(10))/2
               .81465
```
(60)

```
1-Var Stats L1
```
```
1-Var Stats
↑n=36
 minX=.7901
 Q1=.81465
 Med=.8171
 Q3=.8209
 maxX=.8295
```
(61)

3. [pg. 89] Find the 4th decile or 40th percentile ($P_{40}$).

   Since 40% of 36 is 14.4, not a whole number, round up to 15 for the 15th value in the list is $P_{40}$ = 0.8161 lb., as shown in the top lines of screen (62).

```
(40/100)*36
                 14.4
L1(15)
                .8161
L1→CRGWT
{.7901 .8044 .8...
```
(62)

4. Save the data in a list called CRGWT, as in the last lines of screen (62), unless it is already. You will use this data later (page 38) but it will not need to be sorted.

# BOXPLOTS AND FIVE-NUMBER SUMMARY [pg. 96]

**EXAMPLE Qwerty Keyboard Ratings** [pg. 96]: Data stored in list LQWERT on page 15.

(a) Find the values of the five-number summary.

(b) Construct a boxplot.

(a) Press **STAT** <CALC> **1**: 1-Var Stats LQWERT, and then **ENTER**, and use the ▼ key for the second screen of output of screen (63) with minX, $Q_1$, Med, $Q_3$, maxX, or 0, 2, 4, 6, 14, the **five-number summary**.

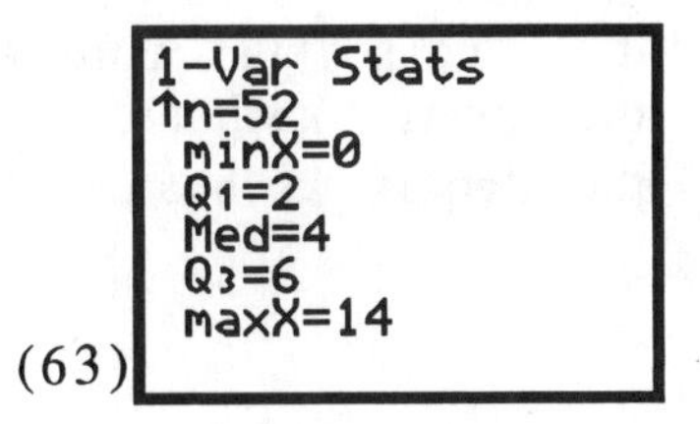

(63)

(b) **Boxplot (or Box-and-Whisker Diagrams)**

1. Using ↑**STAT PLOT**, set up Plot1 as in screen (64). Under "Type," the fifth option (or the second in the second row), is highlighted. Set Plot2 as in screen (65). Under "Type," the fourth option (or the first in the second row), is highlighted — this is for what is called a 'modified boxplot.' Both are shown for comparison but only one need be chosen — my vote is for the 'modified boxplot' as it contains more information as we now explain.

2. Press **ZOOM 9**:ZoomStat for screen (66). The two boxplots differ only in the length of the right whisker.

3. Press **TRACE** for a screen similar to screen (67). The ▼ key can be used to move up and down from the boxplot on top to its 'modified' version on the bottom. You can use the ▶◀ keys to move right and left within a boxplot and to display the **five-number summary** (plus the extra value, X=10, on the modified plot) below the plots, as shown in the last line of screen (67). There are no other values between 10 and maxX = 14, which is considered an **outlier** [pg. 94]. This can be verified with the dotplot on page 17.

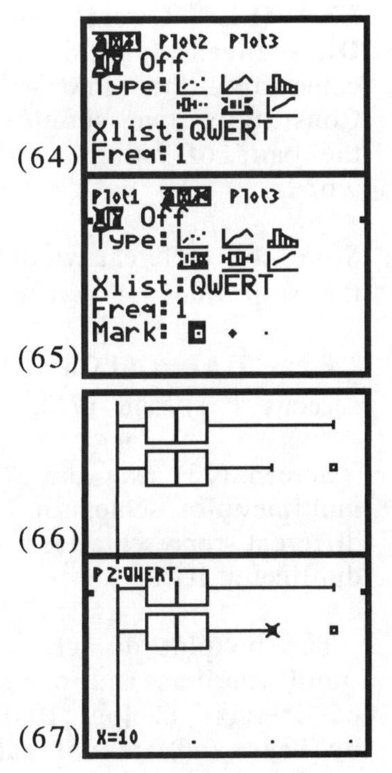

(64) (65) (66) (67)

## Comparing Data Sets with Side-by-Side Boxplots [pg. 97]

**Qwerty and Dvorak Word Rating Comparisons** [pg. 97 Statdisk output]:

Use the data stored in list LQWERT, on page , 15 and store the dvorak keyboard word ratings, repeated at the right, in list LDVORA. (Store row by row as you did for list LQWERT.)

Dvorak Keyboard Word Ratings

| | | | | | | | | | |
|---|---|---|---|---|---|---|---|---|---|
| 2 | 0 | 3 | 1 | 0 | 0 | 0 | 0 | 2 | 0 |
| 4 | 0 | 3 | 4 | 0 | 3 | 3 | 1 | 3 | 5 |
| 4 | 2 | 0 | 5 | 1 | 4 | 0 | 3 | 5 | 0 |
| 2 | 0 | 4 | 1 | 5 | 0 | 4 | 0 | 1 | 3 |
| 0 | 1 | 0 | 3 | 0 | 1 | 2 | 0 | 0 | 0 |
| 1 | 4 | | | | | | | | |

Set Stat Plots 1 and 2 ON as in screen (68).

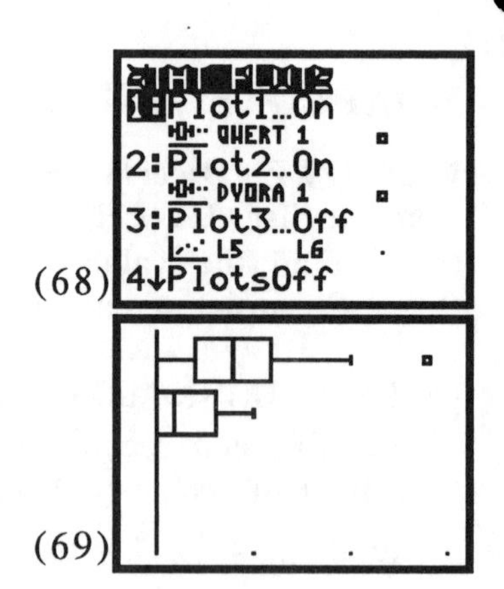


Press **ZOOM** **9**:ZoomStat for screen (69).
In addition to the properties mentioned in the text the dvorak ratings have no outliers. Looking at the raw data, we see the word with the difficult 14 rating on the qwerty keyboard has a much easier 1 rating with the dvorak keyboard. (The sixth word in the fifth row.)

## DIFFERENCE OF TWO LISTS

**EXAMPLE Qwerty and Dvorak Word Rating Differences** [pg 98]: Since the two keyboard ratings came from the same words in the Preamble to the Constitution we should explore the differences between the pairs of ratings corresponding to each of the 52 words.

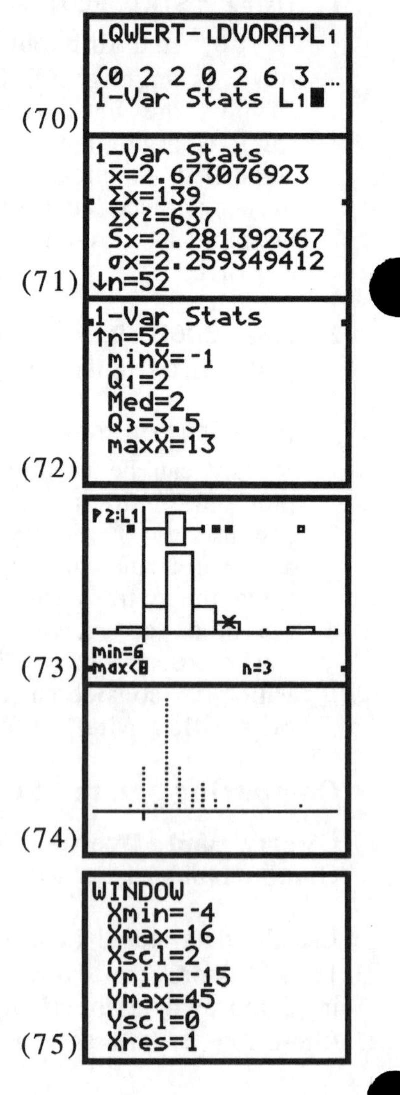


Store the different word ratings in list L1 as shown in the stop lines of screen (70).

Press **STAT** <CALC> **1**: 1-Var Stats L1, and then **ENTER**, for screens (71) and (72).

Screen (73) uses the WINDOW of screen (75) and multiple plot setups of two different "Types" to give different representations of the same data. Can you duplicate it?

The boxplot of screen (73) indicates there are three 'mild' outliers and one 'extreme' outlier, in the notation of Exercise 11 [pg. 105]. (The mild outliers were darkened by hand and are not automatically by the TI-83 Plus.)

The dotplot in screen (74), with screen (73), show that the mild outliers were ratings of -1, two 6s and a 7 with one extreme outlier of 13. (The dotplot is not done automatically on the TI-83 Plus — the one shown was done with a scatterplot, of chapter 9, on a sorted list of the differences and a corresponding list with the count of the differences. For example the first nine values of the first list was -1, 0, 0, 0, 0, 0, 0, 0, 0, 1 and in the second list was 1, 1, 2, 3, 4, 5, 6, 7, 8, 1. Tracing on the boxplot also reveals there are two 6s as it takes two presses of the ▶ key to move from the 5 at the end of the whisker to the 6 and then to the 7 and only one ▶ to move to the 13.

# 3 Probability

In this chapter you will learn some helpful techniques for calculating probabilities, such as using the change-to-fraction function and the raise-to-the-power key, and for calculating factorials, permutations, and combinations. You will also be introduced to simulations, which not only give approximate values for probabilities but also help elucidate probability concepts.

## THE LAW OF LARGE NUMBERS [pg. 116]

**As a procedure is repeated again and again, the relative frequency probability of an event tends to approach the actual probability.**

**EXAMPLE** [Figure 3-2, pg. 116]: The figure in the text which is similar to screen (9)) shows computer-simulated results of the law of large numbers. Notice that when the number of births on the x-axis is small, the proportion of girls greatly fluctuates. As the number of births increases, the proportion of girls approaches 50% or the middle horizontal line. If we flip coins and let H, or a head, represent a girl birth, and T, or tail, represent a boy birth, we can model this situation by flipping many coins. But it is much easier to let the computer do the simulation. The following steps for doing this simulation show some of the possibilities for doing simulations with your calculator. Follow along with your TI-83 Plus.

(1)

1. Put the numbers 1 to 150 in L1 with **↑LIST<OPS>5:seq(X,X,1,150 STO▸ L1**, as in the screen (1).
2. Use **321 STO▸ rand ENTER** to set the seed (as explained on page 4) so that you can follow along. (See the first two lines of screen (2).)
3. Generate one (1) random coin toss at a time with a 0.5 chance of getting a Girl = G = 1. Do this 150 times and store the results in L2 with **MATH<PRB>7:randBin(1,.5,150 STO▸ L2**. The results start with {1 0 1 0 1 1 0 ... and can be read as {G B G B G G B ... .
4. **↑LIST<OPS>6:cumSum(L2 )STO▸ L3 ENTER** stores the cumulative sum of L2 in L3. (See screen (2).). The results {1 1 2 2 3 4 4 ... indicate that the first girl occurred on the first birth (1), the second girl on the third birth (1 + 0 + 1 = 2), the third girl at the fifth birth (1 + 0 + 1 + 0 + 1 = 3), and so on.
5. **L3÷L1 STO▸L4** gives the proportion of heads, or {1 .5 .6666... , indicating 100% (1/1=1) girls on the first toss, 50% (or 1 of 2 heads after the second toss since there is still just one head), and 66.67% heads (or 2 out of 3 heads)

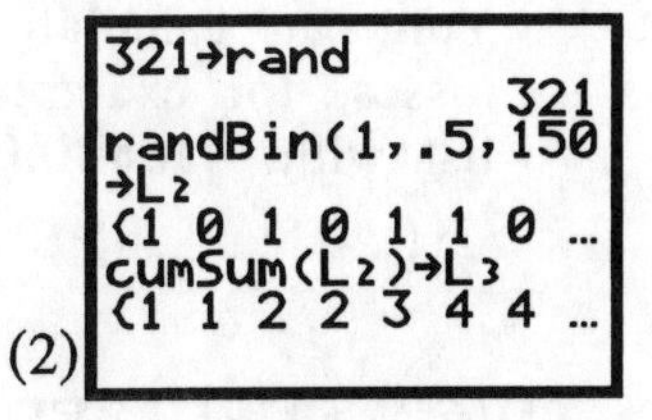

(2)

```
L3
{1 1 2 2 3 4 4 ...
L3/L1→L4
{1 .5 .66666666...
```
(3)

after the third toss. This is a lot of fluctuation in the short run.

6. To see what happens in the long run, set up and plot the results as follows.
   (a) Press **↑STAT PLOT 1:Plot1** to set up for an xyLine plot as in screen (4).
   (b) Press the **Y=** key, and set Y1 = **.5** to plot this horizontal line as in screen (5).
   (c) Set **WINDOW** as shown in screen (6) and then press **TRACE** and **▶▶▶▶▶** for the screen (7) plot of the first 20 tosses since Xmax = 20.
   (d) Change Xmax = 100, leaving the other values as before. (See screen (8).) Press **TRACE** then **▶▶▶▶▶** for the plot of screen (9), which covers 100 tosses. The first tick mark on the X-axis marks off the first 10 tosses, and the plot to that point is the same as in screen (7) but condensed by the change of scale. Notice how the plotted points hover about the Y1 = 0.5 line.

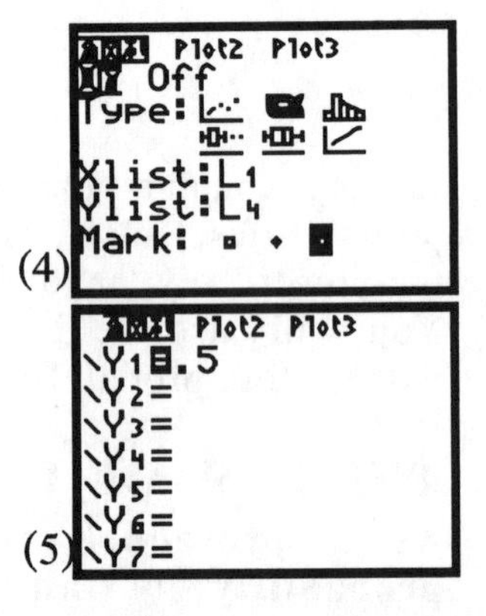


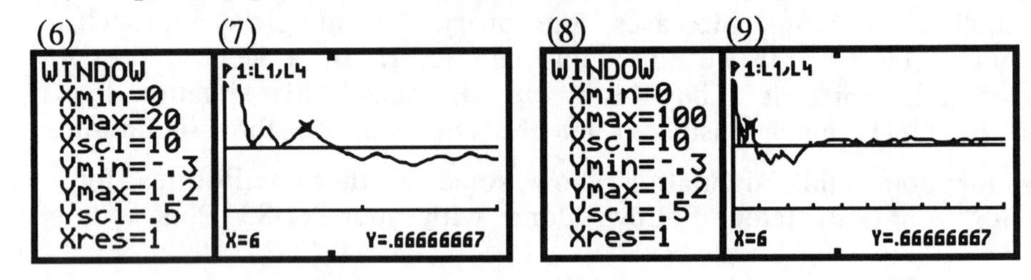


7. You can check intermediate results as shown in screen (10), for example pressing **L4** followed by **(100)** and then **ENTER** gives 0.55 or 55 out of the first 100 births are girls.

   You can calculate a proportion after a given number of tosses, for example 150, without saving anything in a list with **randBin(150,0.5)÷150** as in screen (11).

(10)

```
L4(3)
           .6666666667
L4(100)
                   .55
L4(150)
           .5066666667
```

(11)

```
321→rand
                   321
randBin(150,0.5)
/150
           .5066666667
```

# TWO USEFUL FUNCTIONS

## Change to Fraction Function

**EXAMPLE Women College Graduates** [pg. 119]: If an American woman over age 25 is randomly selected, find the approximate probability that she has a bachelor's degree. A study of 2500 randomly selected women aver age 25 showed that 545 of them have bachelor's degree.

1. Using the above sample we estimated the probability is 545 ÷ 2500 = 0.218. (See

the first lines in screen (12).)

2. Press **MATH 1:▶Frac**, which gives Ans ▶Frac, and then **ENTER**, which gives 109/500 (or 545/2500 reduced), as in the last lines of screen (12).
   **Note**: This change-to-fraction function is handy when you prefer to give the answer as a fraction.

```
545/2500
                .218
Ans▶Frac
             109/500
```
(12)

### The Raise-To-the-Power Key
### Probability of "At Least One" [pg. 144]

**EXAMPLE Gender of Children** [pg. 144]: Find the probability of a couple having at least 1 girl among 3 children.

1. Find the probability of the complement.
   P(boy and boy and boy) = 0.5∗0.5∗0.5 = $0.5^3$
   Type **0.5^3** and then press **ENTER** for 0.125, as in the first two lines of screen (13).
   **Note**: Use the raise-to-the-power key ^ at E5.

2. Type **1 - ↑ANS** and then press **ENTER** for 0.875, the complement of the above and the solution to the problem.
   **Note**: **↑ANS** is at D10.
   This could be changed to a fraction, 7/8, like what was done for screen (12) and shown in screen (13).

```
0.5^3
                .125
1-Ans
                .875
Ans▶Frac
                 7/8
```
(13)

## PROBABILITIES THROUGH SIMULATION [pg. 151]

Finding probabilities of events can sometimes be difficult. We can often benefit from using simulation. You should review the first example in this chapter before doing the examples in this section. Each example assumes you are familiar with the examples that came before it.

**EXAMPLE Gender Selection** [pg. 151]: When testing techniques of gender selection, medical researchers need to know probabilities values of different outcomes, such as the probability of getting at least 60 girls among 100 children. Assuming that male and female births are equally likely, describe a simulation that results in the genders of 100 newborn babies.

One approach is to use the method of screen (2) on page 25 with **MATH<PRB>7:randBin(1,.5,100 STO▶ L1**. The results start with {1 0 1 0 1 1 0 ... and can be read as {G B G B G G B ... .

We use **MATH<PRB>5:randInt(0,1,100 STO▶ L1**. in screen (14) with the integers 0 and 1 being equally likely. (See page 4 for more on 'randInt'.) Press **ENTER** for the results that start with {0 1 0 1 0 0 1... and if we let '0' be Girl and '1' be Boy can be read as {G B G B G G B ... as before. Getting the same result does not surprise us as we used the same seed. That the '0' and '1' were interchanged for the two functions

```
321→rand
                 321
randInt(0,1,100→
L1
{0 1 0 1 0 0 1 ...
sum(L1)
                  45
```
(14)

has to do with how they were constructed.

**↑LIST<MATH>5**:sum(**L1**) **ENTER** sums all the '1s' for 45 boys as in the last line of screen (14) which agrees with the 55 girls out of the first 100 births given in screen (2). These 100 births showed less than 60 girl births. The procedure could be repeated to see how likely it is to get 60 or more girls when we expect about 50.

**EXAMPLE Same Birthdays** [pg. 152]: One classic exercise in probability is the <u>birthday problem</u>, in which we find the probability that in a class of 25 students, at least 2 students have the same birthday. Ignoring leap years, describe a simulation of the experiment that yield birthdays of 25 students in a class.

First set the seed as in the first two lines of screen (15). Generate 25 random integers between 1 and 365, and store these results in L1 then put the data in order with **MATH<PRB> 5**:randInt(1,365,25 **STO▸ L1 ALPHA : STAT 2**:SortA(**L1**). Press **ENTER** for 'Done' in screen (15).

**Note**: The colon (above the decimal key at C10) lets us keep several statements together.

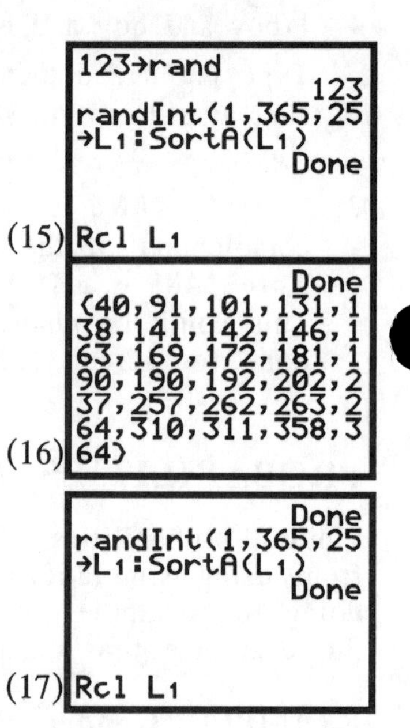

(15) (16) (17)

Recall the values in the sorted list with **↑RCL** L1 then **ENTER** for screen (16) with all the birthdays shown in order. You will discover two 190s, or two students who have the same birthday.

**Note: ↑RCL** (at A9), and discussed on page 7.

Press **↑ENTRY** and **ENTER** then **↑RCL** L1 again for screen (17). Press and check the birthdays of the next class of 25 students (not shown) reveals that no pairs have the same birthday.

Just with these two tries, we have a 50% chance of getting at least two students in the class with the same birthday. But two tries are not enough. Keep repeating the above and investigating the resultant lists for more ties and for a better estimate of the desired probability.

**EXAMPLE Simulating Dice** [pg. 153]: Describe a procedure for simulating the rolling of a pair of dice.

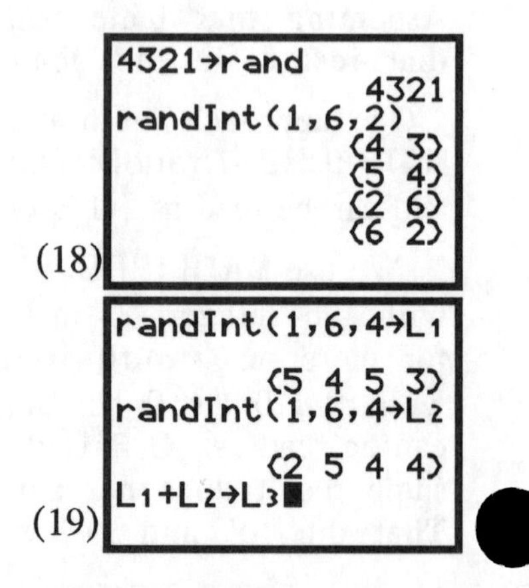

(18) (19)

After setting a random seed, only needed if you want to duplicate these results, generate 2 integers between 1 and 6 for {4 3} as in the 3rd and 4th lines of screen(18). Each press of **ENTER** brings another roll of two dice with sums of 7, 9, 8 and 8 for the four rolls in screen (18).

Another way to simulate four rolls is as in screen (19) with all the results shown with **STAT 1**:Edit in screen (20). The sums of 7, 9, 9, and 7 are given in L3.

If you did such a simulation many times, say 100, you could use a histogram to construct a frequency table of the sums, similar to that shown on page 16 but with a WINDOW having values Xmin = 1.5, Xmax = 12.5 and Xscl = 1. You could use these results to estimate the probability of say getting a sum of 7. The first example above gave 1 of 4 = 25% or 0.25 while the second method gave 0.50. This is a large difference but they were both based on only four rools so we should expect such variation.

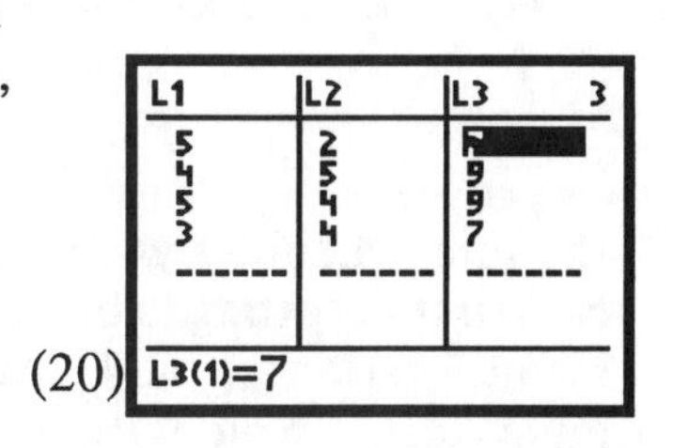

(20)

## COUNTING [pg. 155]

### Factorial!

**Notation: factorial symbol !** [pg. 157]: 4! = 4*3*2*1 = 24

**EXAMPLE Routes to All 50 Capitals** [pg. 160]: How many different routes are possible if you must visit each of the 50 state capitals?

You can visit any one of the 50 states first. For each state, there are 49 ways of picking the second state, 48 ways of picking the third state, and so on until only one state is left to visit. Thus 50*49*48 . . . 3*2*1 = 50!

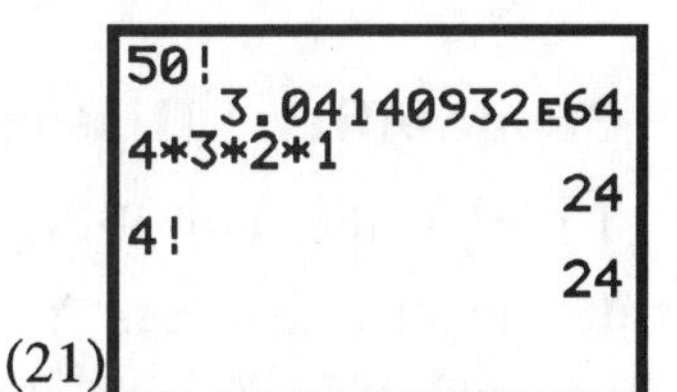

(21)

Type **50** and then press **MATH<PRB>4**:! Then press **ENTER** for the first lines of screen (21) or approximately 3E64, or 3*10^64, or 3 followed by 64 zeros — a lot of possibilities!

### Permutations [pg. 159]

**EXAMPLE Frank Sinatra** [pg 160]: From a list of Sinatra's top 10 songs you must select 3 that will be sung in a medley as a tribute at the next MTV Music Awards ceremony. The order of the songs is important so that they fit together well. How many different sequences are possible?

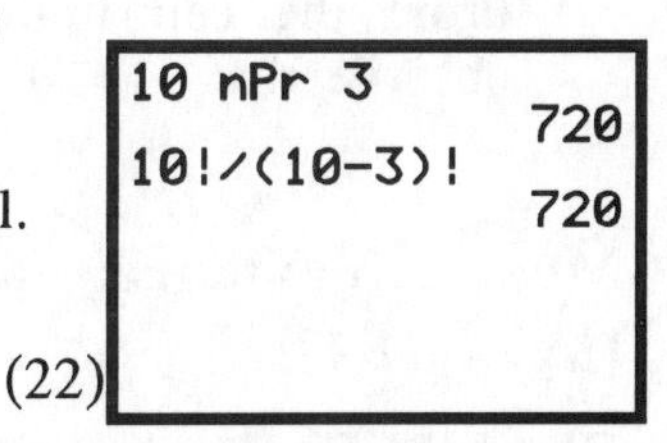

(22)

Type **10**, press **MATH<PRB>2**:nPr, and then type **3**. Press **ENTER** for 720 possible lineups of 3 songs, as in screen (22).

### Combinations [pg 161]

**EXAMPLE** [pg. 161]: The Board of Trustees at the author's college has nine members. Each year, the board elects a three-person committee to oversee buildings and grounds. How many different three-person committees are possible?

9 nCr 3
84
9!/(6!3!)
84

(23)

Type **9**, press **MATH<PRB>3**:nCr, and then type **3**. Press **ENTER** for 84 possible committees, as in screen (23).

# 4 Probability Distributions

In this chapter you will learn about discrete probability distributions by first plotting the probability histograms and then calculating the mean, variance, and standard deviation of a distribution given in table form. You will also find out how to calculate the binomial and Poisson probabilities on the Home screen and then use the ↑DISTR functions, which will alleviate the need to calculate probabilities or to look them up in a table.

| L1<br>x | L2<br>P(x) |
|---|---|
| 0 | 0+ |
| 1 | 0.001 |
| 2 | 0.006 |
| 3 | 0.022 |
| 4 | 0.061 |
| 5 | 0.122 |
| 6 | 0.183 |
| 7 | 0.209 |
| 8 | 0.183 |
| 9 | 0.122 |
| 10 | 0.061 |
| 11 | 0.022 |
| 12 | 0.006 |
| 13 | 0.001 |
| 14 | 0.0+ |

## PROBABILITY DISTRIBUTION BY TABLE [pg. 181]

**EXAMPLE Gender of Children** [pg. 181]: Table 4-1, reproduced at the right, describes the probability distribution for the number of girls among 14 randomly selected newborn babies.

### Probability Histogram [pg. 183 and Figure 4-3]

To plot the probability histogram from the table follow these steps:

1. Put the x values in L1 (you could use seq(X,X,0,14→ L1) and the P(x) values in L2 (0, not 0+, for the first and last values). **Note**: sum L2=0.999 or approximately 1.0.

2. Set up Plot1 and the WINDOW as in screens (1) and (2). Press TRACE for the probability histogram in screen (3). From the cell highlighted and centered at 7, we observe that n = 0.209 = P(7).

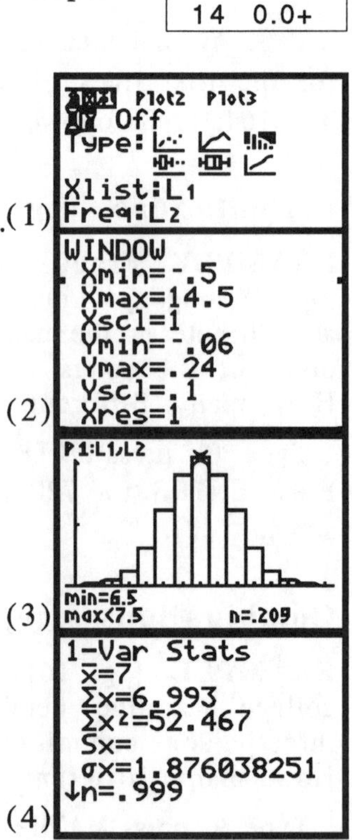


### Mean, Variance, and Standard Deviation [pg. 184]

**EXAMPLE Gender of Children** [pg. 186]: Use the probability distribution to find the mean number of girls (among 14), the variance, and the standard deviation.

1. Put the x values in L1 and the P(x) values in L2, as before.

2. Press STAT<CALC>1:1–Var Stats L1,L2, and then ENTER for screen (4) with $\bar{x}$ actually being $\mu = 7 = \Sigma x \div n$

   $\sigma = \sigma x = 1.876038251$ so $\sigma^2 = 1.87604^2 = 3.5195$.

   **Note**: that n = .999 equals the sum of the probabilities. If n = 1 these results would agree with the next example.

## Expected Value or Mean by Σ[x∗P(x)]

**EXAMPLE Gender of Children** [pg. 187, Table 4-2]: Use the probability distribution to find the expected number of girls or the mean number of girls (among 14).

1. Put the x values in L1 and the P(x) values in L2, as before and partially shown in screen (5).

2. With L3 highlighted multiply L1 by L2 as shown in the bottom line of screen (5). Press **ENTER** for screen (6).

3 Press **↑QUIT**, to return to the home screen, then **↑LIST<MATH>5:sum(L3** and **ENTER** for screen (7) with the sum being E = μ = 6.993 = 7.0 as before.

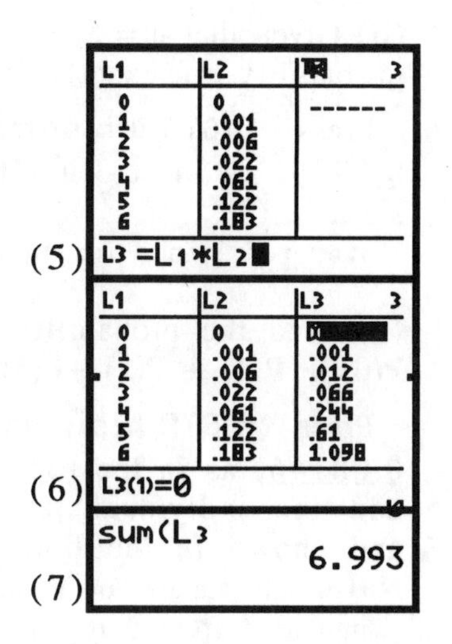

(5) (6) (7)

## BINOMIAL DISTRIBUTION [pg 194]

### Binomial Probability Formula [pg. 196]

$P(x) = nCx * p^x(1 - p)^{n-x}$ for x = 0, 1, 2, . . . , n

**EXAMPLE Using Directory Assistance** [pg. 197]: Find the probability of getting exactly 3 correct responses among 5 different requests from AT&T directory assistance. Assume that in general, AT&T is correct 90% of the time. That is, find P(3) given that n = 5, x = 3, p = 0.9, and q = 0.1.

On the Home screen, type and paste

**5 nCr 3∗0.9^3∗0.1^2**

and then **ENTER** for 0.0729= P(3), as in screen (8).
**Note**: nCr is under **MATH<PRB>**.

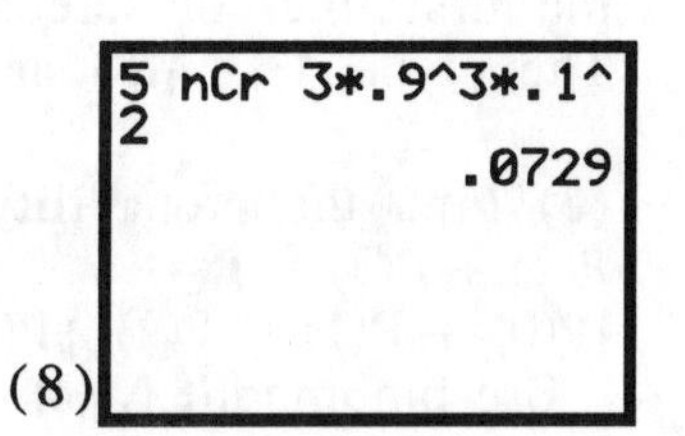

(8)

### ↑DISTR 0:binompdf and ↑DISTR A:binomcdf [pg. 199]

**EXAMPLE Using Directory Assistance**
[pg. 198 (extended to show other possibilities)]:
Given that n = 5, x = 3, p = 0.9, and q = 0.1 find:

(a) the complete probability distribution.

To get the complete table of values, press **↑DISTR 0:binompdf(5,.9 STO▶L2** then **ENTER** for the top lines of screen (9) with P(0) = .00001, P(1) = .00045... Use the ▶ key to reveal the other values or use **↑RCL** L2 as was done for the last lines in screen (9) with P(2) = .0081, P(3) = .0729 and so on. These values could also be seen under L2 in the STAT Editor.

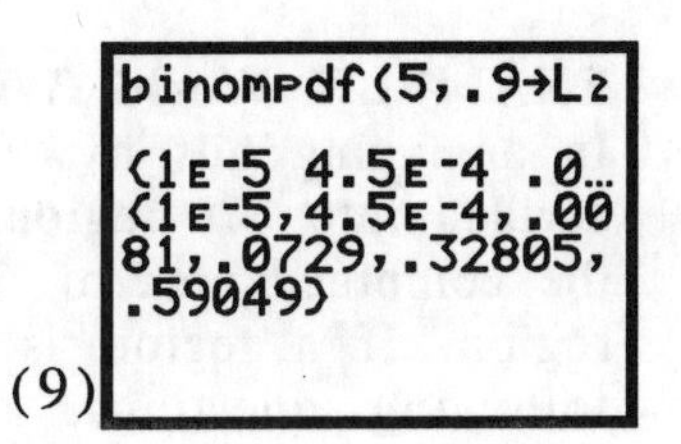

(9)

(**b**) Given that n = 5, x = 3, p = 0.9, and q = 0.1, find the probability of exactly 3 successes.

Press ↑**DISTR 0**:binompdf(**5,0.9,3** and then **ENTER** for 0.0729, as in screen (10). This repeats the solution in screen (8).

**Note**: pdf stands for probability density function.

```
binompdf(5,0.9,3
            .0729
```

(10)

(**c**) Find the probability of at most 3 successes, or P(0) + P(1) + P(2) +P(3).

Press ↑**DISTR A**:binomcdf(**5,0.9,3** and then **ENTER** for 0.08146, as in the top lines of screen (11). You could also add the individual probabilities, given in screen (9) and shown in the bottom lines of screen (11).

**Note**: cdf stands for cumulative density function (cumulative from 0 to x).

Alternatively, you can create a list of P(0) + P(1) + P(2) + P(3), as in the first lines of screen (12), and then sum the values in the list as in the last lines.

**Note**: sum under ↑**LIST** <MATH>.

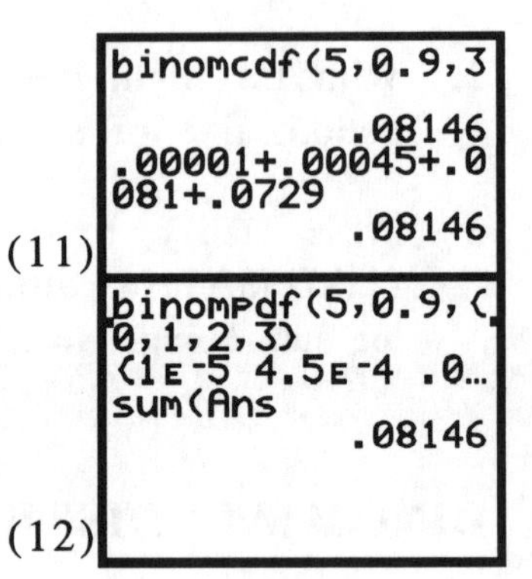


(11)

(12)

(**d**) Find the probability of at least 3 successes, or P(3) + P(4) + P(5) = 1 - [P(0) + P(1) + P(2)].

Press ↑**DISTR A**:binomcdf(**5,0.9,2** and then **ENTER** for the first lines of screen (13) for P(0) + P(1) + P(2). Then type **1** - ↑**ANS** and press **ENTER** for 0.99144.

(**e**) Find the probability of 2 to 4 successes, or P(2) + P(3) + P(4) =
[P(0) + P(1) + P(2) +P(3) + P(4)] - [P(0) + P(1)].

Use binomcdf(**5,0.9,4**) - binomcdf(**5,0.9,1**) = 0..99098 for the solution as in screen (14).

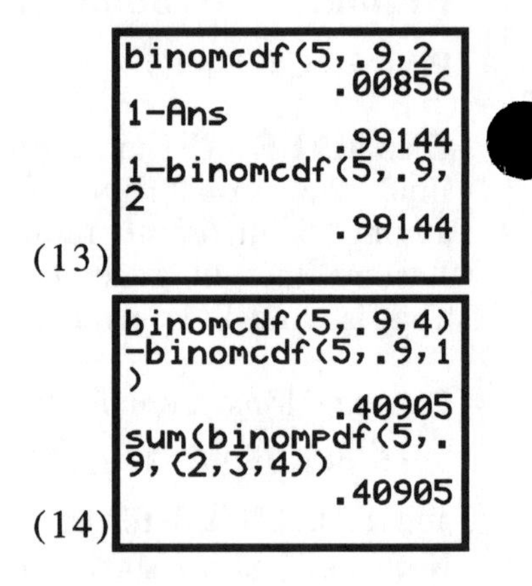


(13)

(14)

## POISSON DISTRIBUTION [pg. 210]

**EXAMPLE World War II Bombs** [pg. 211 (extended to show other possibilities)]: In analyzing hits by V-1 buzz bombs in World War II, South London was subdivided into 576 regions, each with an area of 0.25 sq. km. A total of 535 bombs hit the combined area of 576 regions for an average of $\mu$ = 535/576 = 0.929 hits per region. If a region is randomly selected, use the Poisson distribution to answer the following questions.

**Poisson Probability Formula $P(x) = \mu^x e^\wedge(-\mu)/x!$**

Given $\mu = 0.929$,

(**a**) find the probability that it was hit exactly twice.

(15)

Type in the first line **2→X:0.929^X*e^-0.929/X!** and then press **ENTER** for 0.1704, as in screen (15).

**Note**: We are connecting two statements with a colon (above the decimal point in the last row). We store 2 in X and calculate P(2) = 0.1704. **e^** comes from pressing **↑eˣ** at A8; **!** comes from **MATH<PRB>**. Engaging **↑ENTRY** and pressing ▲ ▲ jumps us to the beginning, where we can type another value for X if we like.

**↑DISTR B:poissonpdf and ↑DISTR C: poissoncdf**

Given $\mu = 0.929$,

(**a**) find the probability that it was hit exactly twice.

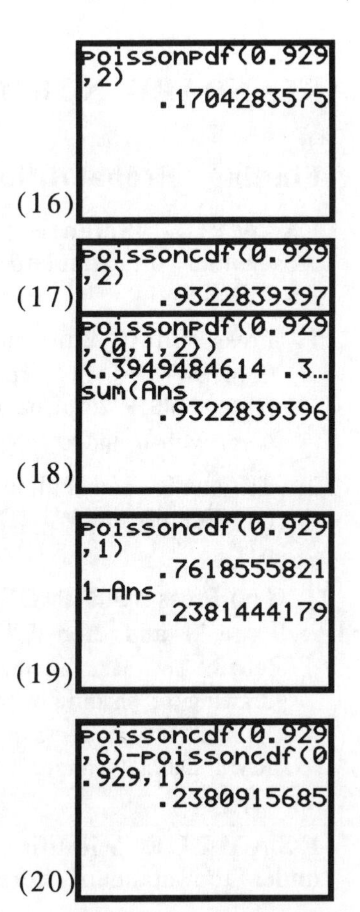

(16) (17) (18) (19) (20)

Press **↑DISTR B:poissonpdf(0.929,2)** and then **ENTER** for screen (16), with P(2) = 0.1704 and as above.

(**b**) find the probability that it was hit at most twice.

Press **↑DISTR C:poissoncdf(0.929,2)** and then **ENTER** for screen (17), with P(0) + P(1) + P(2) = 0.9323.

**Note**: Screen (18) shows another method.

(**c**) find the probability that it was hit at least twice, or P(2) + P(3) + P(4) + P(5)+ ... = 1 - [P(0) + P(1)].

Press **↑DISTR C:poissoncdf(0.929,1)** and then **ENTER** for the first lines of screen (19) for P(0) + P(1). Then type **1 - ↑ANS** and press **ENTER** for 0.23814, which is the solution.

(**d**) find the probability that it was hit from 2 to 6 times. or P(2) + P(3) + P(4) + P(5) + P(6) =
[P(0) + P(1) + P(2) + ... + P(5) + P(6)] - [P(0) + P(1)]
or **↑DISTR C:poissoncdf(0.929,6) - ↑DISTR C:poissoncdf(0.929,1)**
= 0.23809 as in screen (20).

# 5 Normal Probability Distribution

In this chapter you will use the normalcdf and invNorm functions of the TI-83 Plus to calculate areas under a normal curve and to do the inverse, that is, to find a value from a given area or probability. These functions replace the need to use the tables in the text. You will again see the importance of the normal distribution when we simulate the central limit theorem and approximate a binomial distributions with a normal distribution. The normal quantile plot will be introduced to determine whether sample data appear to come from a population that has a normal distribution.

## STANDARD NORMAL DISTRIBUTION [pg. 226]

### Finding Probabilities When Given z Scores [pg. 229]

**EXAMPLE Scientific Thermometers** [pg. 231]: Find the area under the standard normal curve between z = 0 and z = 1.58.

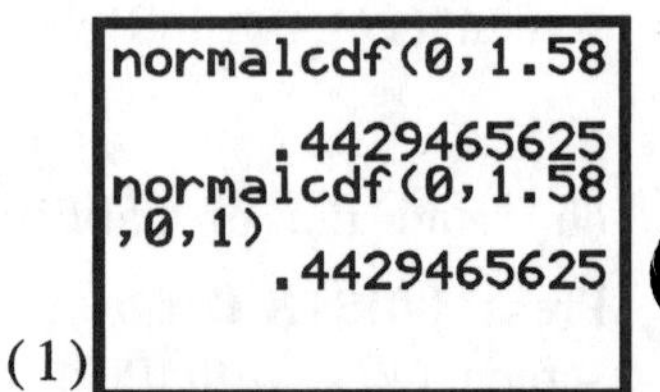

(1)

1. Press **↑DISTR 2**:normalcdf(**0,1.58** and then **ENTER** for 0.442947, as in screen (1).
   **Note**: **↑DISTR 2**:normalcdf(leftmost or low z, rightmost or upper z, μ,σ) with μ and σ optional if they are 0 and 1 respectively.)
2. To shade and calculate the area:
   (a) Set up the WINDOW as in screen (2), with all plots off.
   (b) Press **↑DISTR**<DRAW>**1**:ShadeNorm(**0,1.58** as in screen (3) and then **ENTER** for screen (4).
   **Note**: To shade another area, you will need to clear the previous drawing or shading with **↑DRAW 1**:ClrDraw and then **ENTER** for 'Done.' If you get a mismatch error when trying to shade Reset RAM Defaults under **↑MEM**.

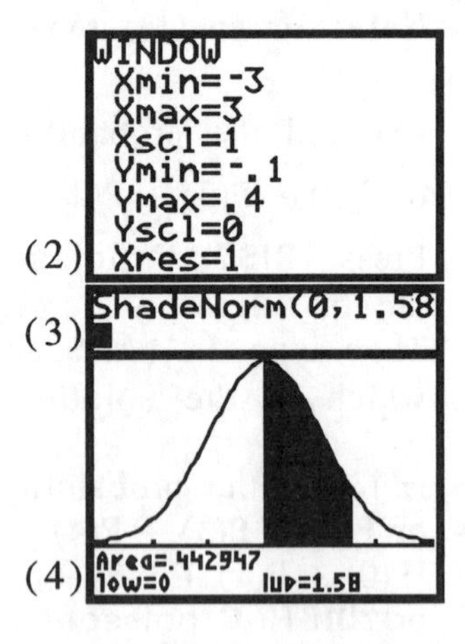

(2) (3) (4)

**EXAMPLE Scientific Thermometers** [pg. 233]: Find the area under the standard normal curve greater than z = 1.27.

1. Press **↑DISTR 2**:normalcdf(**1.27,E99** and then **ENTER** for 0.1020, as in screen (5).
   **Note**: The E for the E99 is from **↑EE** (above the comma) and stands for 10^99, or 10 times itself 99 times, for a very large number to the right. Use **⁻E99** for a value far to the left.

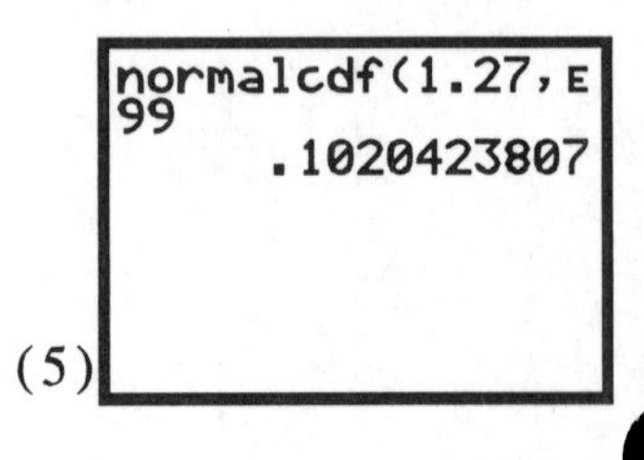

(5)

## Finding Z Scores When Given Probabilities [pg. 235]

**EXAMPLE Scientific Thermometers** [pg. 237]: Find the z score that is the 95th percentile,separating the top 5% from the bottom 95% of the area under the standard normal distribution.

1. Press **↑DISTR 3**:invNorm(**0.95** and then **ENTER** for 1.64485, as in the first two lines of screen (6).
   **Note**: **↑DISTR 3**:invNorm(area to left of z).

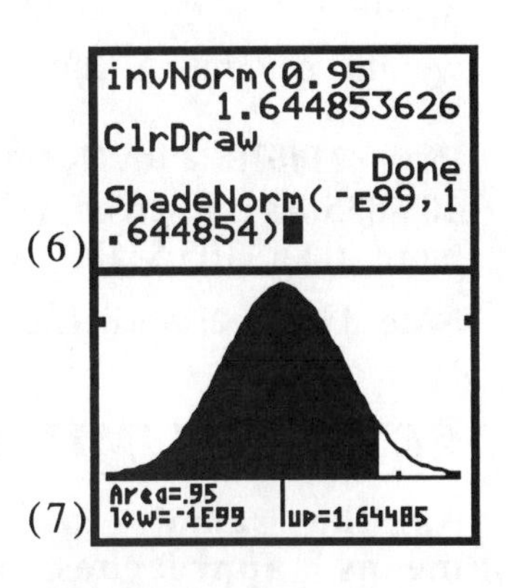


(6)

(7)

2. To shade the desired area:
   (a) Set up the **WINDOW** as in screen (2), with all plots off.
   (b) Press **↑DRAW 1**:ClrDraw and then **ENTER** for 'Done' (as in the middle lines of screen (6)) If you need to clear the shading from a previous problem.
   (c) Press **↑DISTR<DRAW>1**:ShadeNorm(**¯E99,1.644854** for the last two lines of screen (6) and then press **ENTER** for screen (7).

## NORMAL DISTRIBUTIONS Finding Probabilities [pg. 241]

**EXAMPLE Jet Ejection Seats** [pg. 242]: If women's weights are normally distributed with a mean $\mu$ = 143 lb and a standard deviation $\sigma$= 29 lb what is the probability that if a woman is randomly selected that she weighs between 143 lb and 201 lb?

1. Press **↑DISTR 2**:normalcdf(**143,201,143,29** and then **ENTER** for 0.4772499, as in screen (8).
   **Note**: **↑DISTR 2**:normalcdf(low value,upper value,$\mu$,$\sigma$).

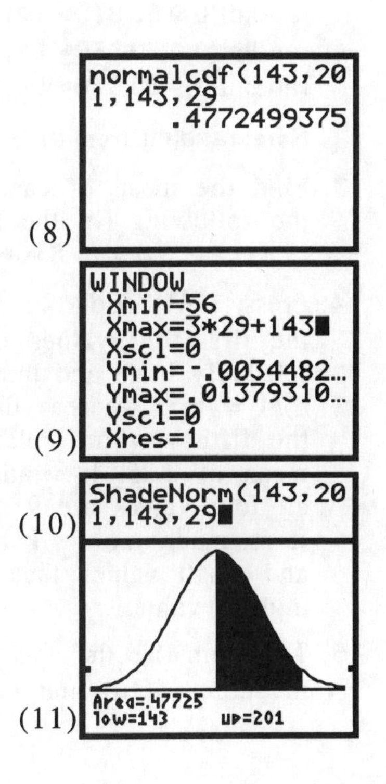


(8)

(9)

(10)

(11)

2. To shade and calculate the area:
   a. Set up the **WINDOW** as in screen (9) as follows with all plots off:
      Xmin = -3*$\sigma$ + $\mu$ = -3*29+ 143 = 56
      Xmax = 3*$\sigma$ + $\mu$ = 3*29 + 143 = 230
      Xscl = 0 and Yscl = 0
      Ymin = -0.1/$\sigma$ = -0.1/29 = -.00345
      Ymax = 0.4/$\sigma$ = 0.4/29 = .01379
      **Note**: Screen (9) shows that the above calculations can be done in the **WINDOW** editor, but the calculation for Xmax shown will be calculated to 230 when you press **ENTER** or use the Down arrow.
   b. Press **↑DISTR<DRAW>1**:ShadeNorm(**143,201,143,29** as in screen (10) and then **ENTER** for screen (11).

   **Note**: To shade another area using the same window, you will need to clear the previous drawing or shading with **↑DRAW 1**:ClrDraw and then **ENTER** for 'Done.'

## NORMAL DISTRIBUTIONS Finding Values [pg. 249]

**EXAMPLE Womens Weights** [pg. 250]: Find $P_{10}$ (the 10th percentile) of women's weights if the weights are normally distributed with $\mu$ = 143 lb and $\sigma$ = 29 lb. That is , find the weight separating the bottom 10% from the top 90%.

Press **↑DISTR 3**:invNorm(**0.10,143,29** and then **ENTER** for 105.835 as in screen (12). Only 10% of women weigh more than 105.8 lb.

**Note: ↑DISTR 3**:invNorm(area to left of score,$\mu$,$\sigma$)

```
invNorm(0.10,143
,29
        105.8350046
```

(12)

## CENTRAL LIMIT THEOREM [pg. 255]

**As the sample size increases, the sampling distribution of sample means approaches a normal distribution.**

[pg. 258]: Here you will see the plausibility of the central limit theorem through a simulation that finds the means of the last four digits of social security numbers.

1. On the Home screen, set your seed so that you can duplicate your results, with **4321 STO▸ rand**, as on page 4 (step 4) and the first two lines of screen (13).
2. Set up the following list as partly shown in screen (13). (Do not forget the last entry feature of **↑ENTRY**, as shown on pages 5 and 6.)
   randInt(0,9,50 **STO▸** L1) **ENTER**
   randInt(0,9,50 **STO▸** L2) **ENTER**
   randInt(0,9,50 **STO▸** L3) **ENTER**
   randInt(0,9,50 **STO▸** L4) **ENTER**

   **Note:** randInt( from **MATH<PRB>** as on page 4.

```
4321→rand
               4321
randInt(0,9,50)→
L1
{5 3 7 5 1 9 9 ...
randInt(0,9,50)→
L2
```

(13)

3. Find the mean of each row of these lists using the following (in the last lines of screen (14)):
   ( L1+L2+L3+L4)/4 **STO▸** L5 **ENTER**

```
{6 3 4 0 5 7 1 ...
randInt(0,9,50)→
L4
{6 0 2 4 3 9 2 ...
(L1+L2+L3+L4)/4→
L5
{6.5 3 5.5 2.5 ...
```

(14)

4. Press **STAT 1**:Edit for screen (15). Notice that the first four values in the first row are relatively large and their mean = (5 + 9 + 6 + 6)/4 = 6.5, whereas the first four values in the fifth row are relatively small with a mean of 2.25. It would be very unlikely that all four values are 9 or all four values are 0. It is much more likely that there are large and small values that will average out to a middle value.

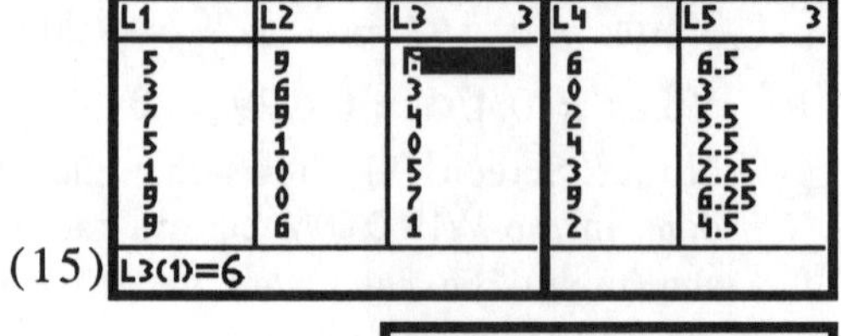

| L1 | L2 | L3 | L4 | L5 |
|---|---|---|---|---|
| 5 | 9 | 6 | 6 | 6.5 |
| 3 | 6 | 3 | 0 | 3 |
| 7 | 9 | 4 | 2 | 5.5 |
| 5 | 1 | 0 | 4 | 2.5 |
| 1 | 0 | 5 | 3 | 2.25 |
| 9 | 0 | 7 | 9 | 6.25 |
| 9 | 6 | 1 | 2 | 4.5 |

L3(1)=6

(15)

(16)

5. To summarize the data, set up the WINDOW and Plot1 as in screens (16) and (17). Press **TRACE** for the histogram

of screen (18). The distribution of the 50 values in L1 is fairly uniform, with about 5 of each of the 10 digits 0, 1, 2, 3, 4, 5, 6, 7, 8, and 9.

6. Set up Plot1 for a histogram of the means stored in L5. **TRACE** reveals the histogram of screen (19). Notice that this histogram is much more normally shaped, with no means of 0 (the first cell) or 9 (the last cell), as we had conjectured. Most of the means are lumped near the center at 4, 5, and 6 and predicted by the central limit theorem.

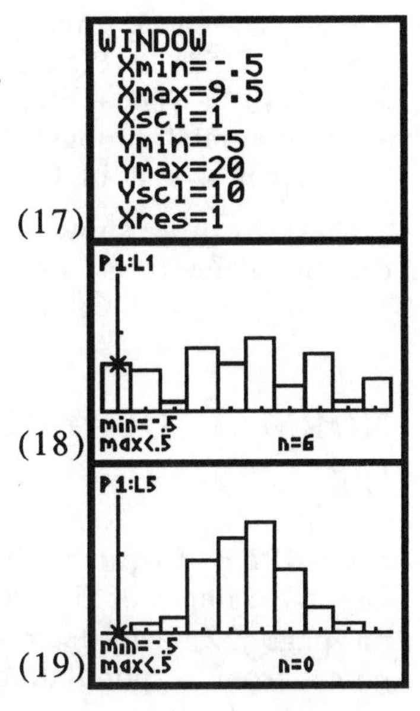

(17) (18) (19)

**EXAMPLE Jet Ejection Seats** [pg. 259]: Given that the population of women has normally distributed weights, with a mean of 143 lbs and a standard deviation of 29 lbs, find the probability that:

(a) if 1 woman is randomly selected, her weight is greater than 150 lbs.
(Answer: 0.405 as in first three lines of screen (20).)

(b) if 36 different women are randomly selected, their mean weight is greater than 150 lbs.
(Answer: 0.074 as in the last three lines of screen (20) using a standard error of 29/√(36).

(20)

It is much less likely that the mean of 36 values is 7 lbs. above the population mean than one randomly selected value.

## NORMAL DISTRIBUTION AS AN APPROXIMATION TO THE BINOMIAL DISTRIBUTION [pg. 267]

**If np ≥ 5 and n(1 - p) ≥ 5, then the binomial random variable is approximately normally distributed with the mean and standard deviation given as $\mu = np$ and $\sigma = \sqrt{(np(1-p))}$**

**EXAMPLE Gender Discrimination** [pg. 270]: Assume that a college has an equal number of qualified male and female applicants, and assume that 520 of the last 1000 newly accepted students are men. Use the normal distribution to estimate the probability of getting at least 520 men if each acceptance is done independently with no gender discrimination.

With n * p = 1000 * 0.5 = 500 > 5 and n * q = 1000*(1 - 0.50) = 500 > 5, we could use the normal distribution to approximate the binomial with $\mu = np = 500$ and $\sigma = \sqrt{(1000*0.5*0.5)} = 15.8114$.

Using the continuity correction to get the area for P(520) + P(521) + P(522) + . . . + P(1000), we need to start with 519.5 and continue out to the right for a normal approximation of 0.10873 as in the first lines of screen (21).

You can check this answer since
P(520) + P(521) + P(522) + . . . + P(1000)
= 1 - [P(0) + P(1) + P(2) + . . . + P(518) + P(519)] and the exact binomial probability can be calculated as we did in Chapter 4 and in the last lines of screen (21) for 0.10872.

These results show how good the normal approximation can be when it is used in the following chapters.

```
normalcdf(519.5,
E99,500,15.8114
            .1087343477
1-binomcdf(1000,
0.5,519)
            .1087241367
```

(21)

## NORMAL QUANTILE PLOTS (for Determining Normality) [pg. 279]

**EXAMPLE Regular Coke** [pg. 281 but text uses Diet Pepsi]: Use the sample of 36 weights of regular coke, saved as LCRGWT on page 22, and determine whether the sample appears to come from a population with a normal distribution.

Set up Plot1 as in screen (22) with 'Type' the last option and Data Axis:X. Press **ZOOM9**:ZoomStat for screen (23).

All but one point, the smallest weight 0.7901 lb which is an outlier, appear to be close to a straight line, suggesting that the sample data come from a population with a normal distribution. If there were more data points filling the gap this would indicate a distribution skewed to the left.

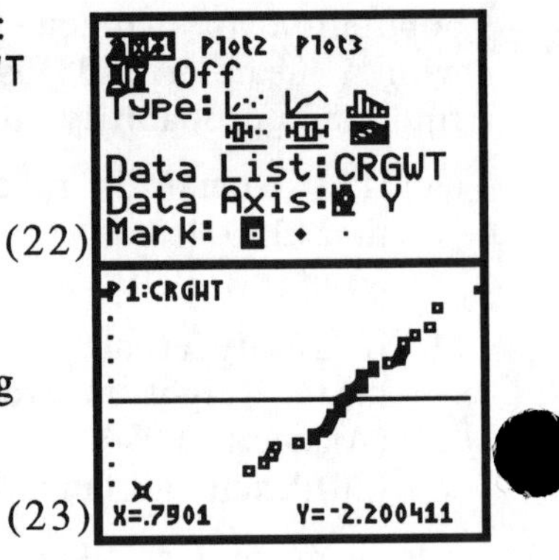


(22)

(23)

## Generating Random Values from a Normal Distribution

Below we mathematically generate random value from a normal distribution with $\mu = 0$ and $\sigma = 1$ (standard normal) using. **MATH** <PRB>**6**:randNorm. Screen (24) shows 5 values being stored in L1 and L2 and their normal quantile plots (as done above) in screen (25) and (26).

These plots are helpful for gaining experience in the type of distributions we should expect when determining normality from sample data of a given sample size. Try some different sizes for yourself.

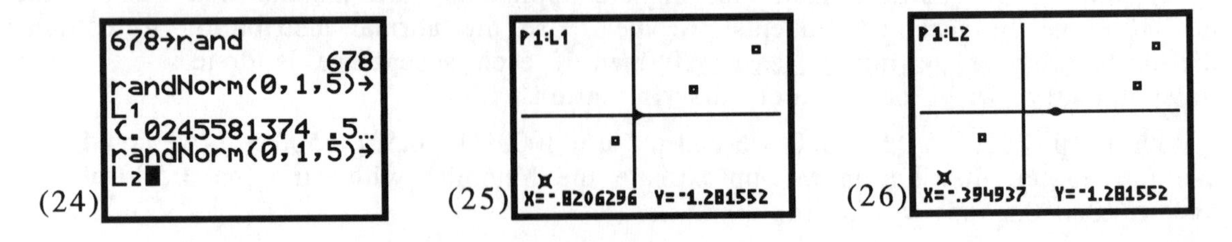


(24) (25) (26)

# 6 Estimates and Sample Sizes

In this chapter you will learn how to estimate population parameters by taking one simple random sample from the population. The **STAT<TESTS>** menu of the TI-83 Plus will be used, but the calculations will also be shown on the Home screen to clarify what equation is being used. The critical values are calculated so there is no need to look them up in tables. To use the **STAT<TESTS>** functions, you can input the data in two ways:

(1) Using summary statistics such as $\bar{x}$ and n (see screens (2) and (12))
(2) Using the raw data stored in a list (see screens (7) and (15)).

## ESTIMATING A POPULATION MEAN: LARGE SAMPLES [pg. 296]

**EXAMPLE Critical Value** [pg. 300]: Find the critical value $z_{\alpha/2}$ corresponding to a 95% degree of confidence.

$\alpha = 1 - 0.95 = 0.05$, $\alpha/2 = 0.05/2 = 0.025$, $1 - 0.025 = .975$.

$z_{\alpha/2} = 1.96$ the value that separates the the top 2.5% area of the standard normal distribution as in screen (1).

**Note:** ↑**DISTR 3**:invNorm(area to the left of z)

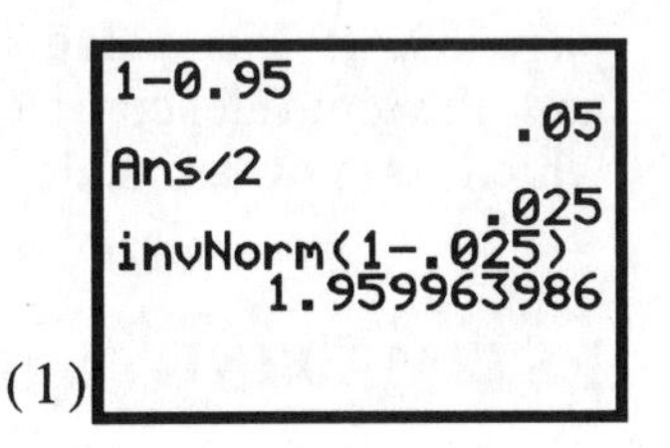

(1)

### Using Summary Statistics

**EXAMPLE Body Temperatures** [pg. 304]:
For the body temperatures in [Table 6-1], we have n = 106, $\bar{x}$ = 98.20° F, and s = 0.62° F. For a 0.95 degree of confidence, use these statistics to find the margin of error, E, and the confidence interval for μ.

ZInterval
Inpt:Data Stats
σ:.62
x̄:98.2
n:106
C-Level:.95
Calculate

(2)

1. Press **STAT<TESTS> 7**:ZInterval for a screen similar to screen (2). Make sure that the input line (Inpt:) has Stats highlighted by using the ▶ key and pressing **ENTER**. Since n = 106 >30, use ▼ and let σ = s = 0.62. Also, input $\bar{x}$, n, and the confidence level (C-Level) as in screen (2).
2. Highlight Calculate in the last line of screen (2) and press **ENTER** for screen (3) with the 95% confidence interval of (98.082,98.318), or 98.08 < μ < 98.32. You can calculate the margin of error by taking the larger value of the confidence interval and subtracting $\bar{x}$, or E = 98.318 − 98.2 = 0.118 = 0.12. So you can also express the 95% confidence interval as 98.2 ± 0.12.

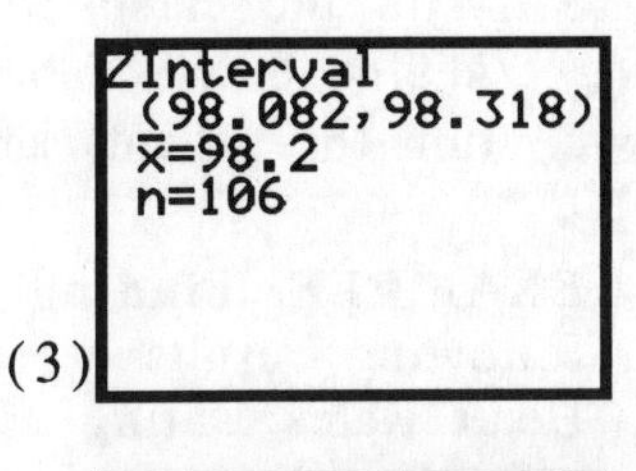

(3)

3. You can do the Home screen calculations using $E = z_{\alpha/2} * \sigma \div \sqrt{n}$ = 1.96*0.62/√(106) **STO▶** E as shown in screen (4).

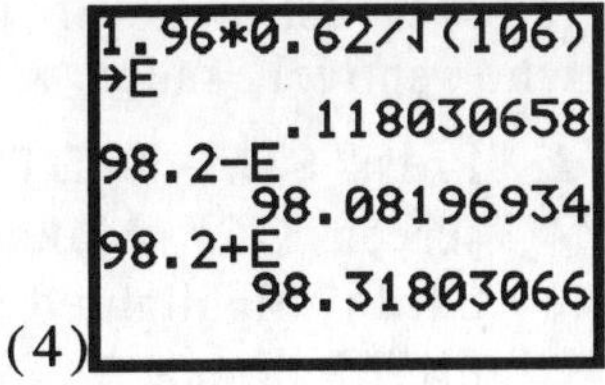

(4)

### Using Raw Data List

**EXERCISE 21 Regular Coke** [pg. 311]: Data on page 22 of the weights of 36 cans of regular coke stored in LCRGWT and last used on page 38. Find the 95% confidence interval for $\mu$.

This is similar to the previous example but first we need to do a 1-Var Stats on LCRGWT to find Sx = .0075 (to estimate $\sigma$) as in screens (5) and (6).

Make sure that the input line (Inpt:) for the ZInterval has Data highlighted as in screen (7) for the output in screen (8) for $0.81437 < \mu < 0.81927$.

Since we did not know $\sigma$ and had to calculate it we could have just as easy used the Stats input of the last example.

There is the complication of the outlier mentioned on page 39. If this value is dropped from the data set Sx = .006 and the confidence interval becomes $0.8156 < \mu < 0.81957$ which is not as wide as the previous interval.

```
1-Var Stats LCRG
WT
```
(5)

```
1-Var Stats
 x̄=.8168222222
 Σx=29.4056
 Σx²=24.0211202
 Sx=.007507372
 σx=.0074023687
↓n=36
```
(6)

```
ZInterval
 Inpt:Data Stats
 σ:.0075
 List:CRGWT
 Freq:1
 C-Level:.95
 Calculate
```
(7)

```
ZInterval
 (.81437,.81927)
 x̄=.8168222222
 Sx=.007507372
 n=36
```
(8)

## ESTIMATING A POPULATION MEAN: SMALL SAMPLES [pg. 312]

**EXAMPLE Finding a Critical value** [pg. 314]: A sample of size n = 15 (df =15-1= 14) is a simple random sample selected from a normally distributed population. Find the critical value $t_{\alpha/2}$ corresponding to a 95% degree of confidence.

1. Press **MATH** 0:Solver for a screen similar to screen (9). (If the screen looks more like screen (10), press ▲.) To get the screen to look like screen (9), paste tcdf( from ↑**DISTR 5**. You will solve for an X that has 0.025 of area to its right under a t-distribution with df = 14.
2. Press **ENTER** for screen (10). Type **2** as first guess for X=2 since you want to be in the right tail.
3. With the cursor flashing on the X=2 line, press **ALPHA SOLVE** (above **ENTER**), and wait for the calculation (be patient) and X = $t_{\alpha/2}$ = 2.145, as in screen (11).

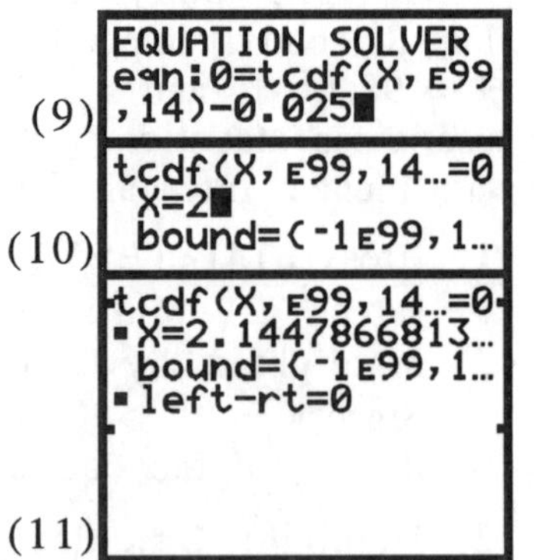


**EXAMPLE Finding a Confidence Interval Using Stats** [pg. 312]: The following results were obtained for an experiment on cardiac demands. Maximum Heart Rates During Manual Snow Shoveling: $\bar{x}$ = 175, s = 15 and n = 10. Find the 95% interval estimate of $\mu$, the mean maximum heart rates (beats per minute) for those who shovel snow.

1. Press **STAT**<TESTS> **8**:TInterval for a screen similar to screen (12). Make sure that the input line (Inpt:) has Stats highlighted by using the ▶ key and pressing **ENTER**. Input $\bar{x}$, Sx, n, and the C-Level as in screen (12).

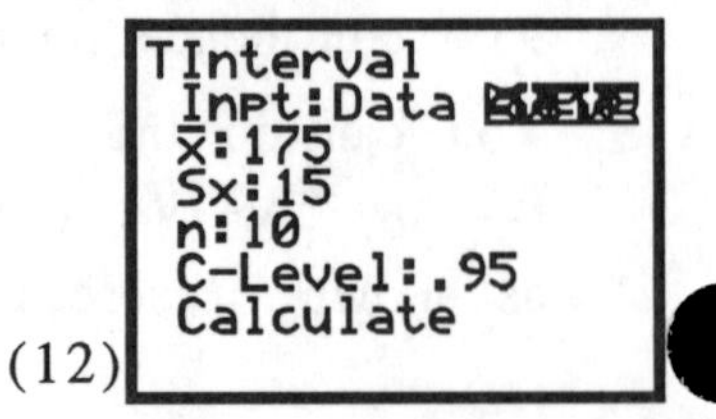

2. Highlight Calculate in the last line of screen (12), and press **ENTER** for screen (13) with the 95% confidence interval of (164.27, 185.73), or $164.27 < \mu < 185.73$. You can calculate the margin of error by taking the larger value of the confidence interval and subtracting $\bar{x}$, or $E = 185.73 - 175 = 10.73$. So you can also express the 95% confidence interval as $175 \pm 10.73$.

```
TInterval
 (164.27,185.73)
 x̄=175
 Sx=15
 n=10
```
(13)

3. You can do the Home screen calculations using $E = t_{\alpha/2} * s \div \sqrt{n} = 2.262*15/\sqrt{(10)}$ **STO▸** E as shown in screen (14).

```
2.262*15/√(10)→E
          10.7296081
175-E
         164.2703919
175+E
         185.7296081
```
(14)

**EXAMPLE Finding a Confidence Interval Using Data**: Data on page 22 of the weights of 36 cans of regular coke stored in LCRGWT and last used on page 40. Find the 95% confidence interval for $\mu$.

1. Press **STAT**<TESTS> **8**:TInterval for a screen similar to screen(15). Make sure that the input line (Inpt:) has Data highlighted, input list LCRGWT with Freq set at 1 and C-Level at 0.95.

```
TInterval
 Inpt:Data Stats
 List:CRGWT
 Freq:1
 C-Level:.95
 Calculate
```
(15)

2. Highlight Calculate and press **ENTER** for screen (16). The 95% confidence interval is (.81428, .81936) or $0.81428 < \mu < 0.81936$. Compared this to screen (8) with $0.81437 < \mu < 0.81927$ which is basically the same because the t and z critical values are essentially the same for large degrees of freedom. I think the TInterval is easier for raw data no matter the sample size if $\sigma$ is unknown as the function calculates and uses Sx and $\bar{x}$ and n.

```
TInterval
 (.81428,.81936)
 x̄=.8168222222
 Sx=.007507372
 n=36
```
(16)

**Note: Sample Size** The TI-83 Plus does not have a function for determining sample size but calculations can be done on the home screen. An example for proportions is given on page 42.

# ESTIMATING A POPULATION PROPORTION [pg. 329]

**EXAMPLE Misleading Survey Responses** [pg. 332]: In a survey of 1002 people, 701 people said that they voted in a recent presidential election. Find a 95% confidence interval estimate of the population proportion of people who say that they voted. (Notice that $701 > 5$ and $(1002 - 701) > 5$)

1. Press **STAT**<TESTS> **A**:1-PropZInt and then set up a screen similar to the one in screen (17). Input x (the number who said they voted), n (the number in the sample), and the confidence level (C-Level) as in screen (17).

```
1-PropZInt
 x:701
 n:1002
 C-Level:.95
 Calculate
```
(17)

2. Highlight **Calculate** in the last line of screen (17), and press **ENTER** for screen (18), with a point estimate of $\hat{p}$ = 0.6996 = 70% and a 95% confidence interval of (0.6712, 0.7280), or 67.1% < p < 72.8%. You can calculate the margin of error by taking the larger value of the confidence interval and subtracting $\hat{p}$, or E = .7280 − .6996 = 0.0284. So you can also express the 95% confidence interval as 0.70 ± 0.028.

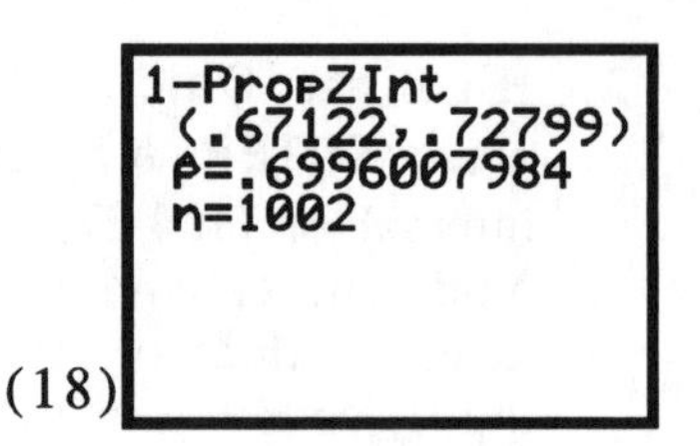

(18)

3. The home screen calculations are shown in screens (19) and (20) using the following formulas:
$\hat{p}$ = 701/1002 = 0.6996 = P
$E = z_{\alpha/2} * \sqrt{(P*(1-P)/n)}$
$= 1.96 * \sqrt{(0.6996*(1-0.6996)/1002)} = 0.028$

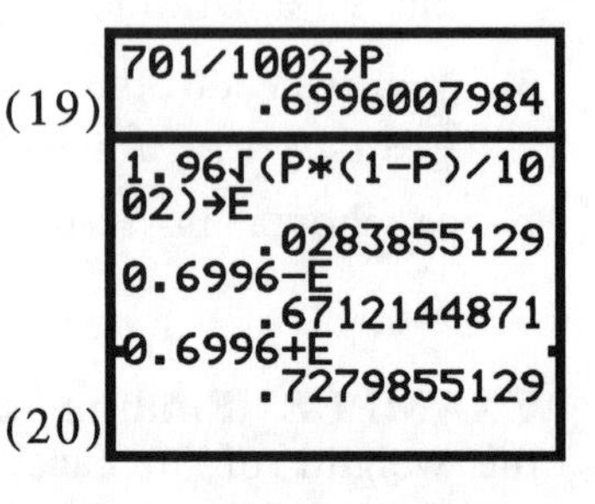

(19) (20)

**Note**: If $\hat{p}$ were given as 0.6996 and n = 1002, you could not input x:0.6996*1002 for 700.9992 in screen (17). Doing so would cause a domain error since x must be an integer. You must first round to the nearest integer, or 701.

## Determining Sample Size [pg. 334]

**EXAMPLE E-Mail** [pg. 335] How many households must be surveyed in order to be 90% confident that the sample percentage is in error by no more than four percentage points (E = .04)?
**a**. If an earlier study (1997) indicated 16.9%.
**b**. Assume we have no prior information on p.

Use Formula $n = (z_{\alpha/2})^2 \hat{p}(1 - \hat{p})E^2$ [pg. 334].

Find the critical value as in screen (1) and in screen (21), with α = 1.0 − .90 = .10, for $z_{\alpha/2}$ = 1.645

Using the above with $\hat{p}$ = 0.169 we get a sample size of about 238, while if we had no idea of p the size would be 423 as in screen (22) using $\hat{p}$ = 0.5.

```
invNorm(1-.10/2)
         1.644853626
```
(21)

```
1.645²*.169*(1-0
.169)/.04²
         237.5196531
1.645²*.5*(1-.5)
/.04²
         422.8164063
```
(22)

# ESTIMATING A POPULATION VARIANCE [pg. 342]

**EXAMPLE Critical Values for Body Tempertures** [pg. 345]: Table 6-1 data is approximately bell-shaped. with mean 98.6° F , s = 0.62° F, and n = 106. There are no outliers. Construct a 95% confidence interval estimate of σ, the standard deviation of the body temperature of the whole population.

With degrees of freedom = 106 − 1 = 105 we can calculate these critical values with the **MATH** Solver, similar to what was done for the critical t-value in screens (9), (10), and (11). For example screens (23), (24), and (25) are for calculating

$\chi^2_L$ =78.536, and screens (26), (27), and (28) for calculating $\chi^2_R$ = 135.247.

Using $(n - 1)s^2/\chi_R^2 < \sigma^2 < (n - 1)s^2/\chi_L^2$ [Formula on pg. 346]

$$(106 - 1)0.62^2/135.247 < \sigma^2 < (106 - 1)0.62^2/78.536$$
$$0.2984 < \sigma^2 < 0.5139$$
$$0.55 < \sigma < 0.72.$$

See screen (29) for the lower limit calculations. The confidence interval becomes 0.55° F < σ < 0.72° F, which is close to the results obtained in the text with the critical values approximated.

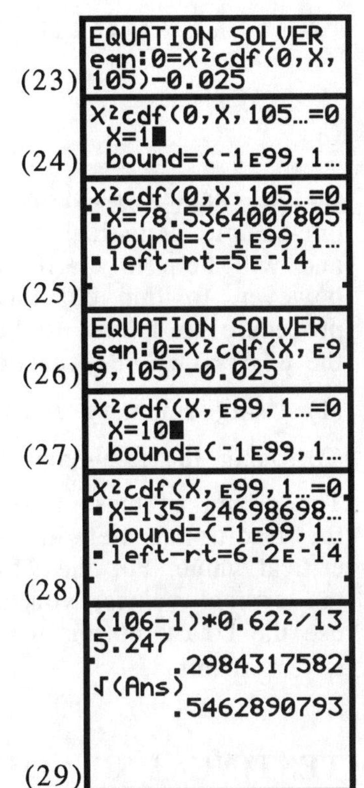


## TECHNOLOGY PROJECT (THE BOOTSTRAP METHOD FOR CONFIDENCE INTERVALS) [pg. 358]

L3 = {2.9 564.2 1.4 4.7 67.6 4.8 51.3 3.6 18.0 3.6}

Using the sample data given above, and storing it in L3, construct a 95% confidence interval estimate of the population mean μ by using the following method.

(a) Create 500 new samples, each of size 10, by selecting 10 scores with replacement from the 10 sample scores above.

(b) Take the mean of each sample.

(c) Put the sample means in order.

You can carry out steps (a), (b), and (c) on the TI-83 Plus with the program listed in screens (30) and (31).

Screens (32) and (33) show the output of running the program after setting a seed.

(d) Calculate $P_{2.5}$ and $P_{97.5}$ as in screen (34) for the estimate $P_{2.5} < \mu < P_{97.5}$ = 6.475 < μ < 184.73.

To calculate the 95% confidence interval estimate for the population standard deviation σ, change **mean** in the program to **stdDev**, which is found under **↑LIST<MATH>7**.

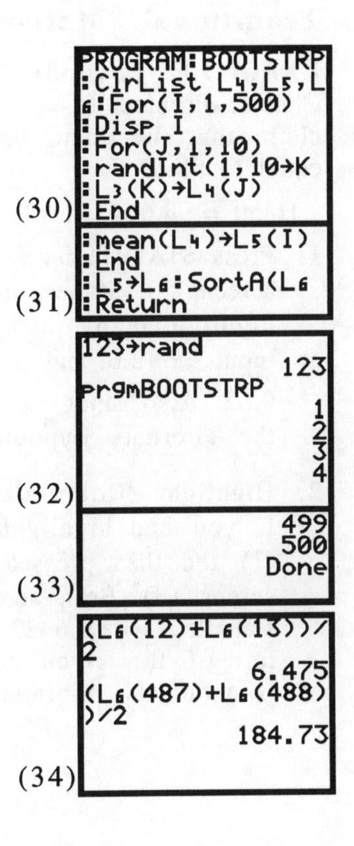

# 7 Hypothesis Testing

In this chapter you'll use the **STAT<TESTS>** functions of the TI-83 Plus to do one-sample statistical tests. Calculations will also be shown, on the Home screen, and will help to clarify the procedure used. You should refer to the main text, however, to find out the requirements necessary to make the test valid and the procedure to follow to include all the steps your instructor has asked for. To use the **STAT<TESTS>** functions, you can input the data in two ways:

(1) Using summary statistics such as $\bar{x}$ and n (see screen (1))

(2) Using the raw data stored in a list (see screen (10))

The output of these tests gives the p-value but also the test statistic, so you can use the traditional method of hypothesis testing, checking the test statistic against the critical value. For the TI-83 Plus, you don't need to use the tables in the text to find the critical values, you can use the procedure introduced in chapter 6. You can use the Draw option for the output (see screens (3) and (5)), which can help clarify a test result.

## TESTING A CLAIM ABOUT A MEAN: LARGE SAMPLES [pg. 381]

### Traditional Method (Two-Tail Test)

**EXAMPLE 1a Body Temperature** [pg. 383]: Using the sample data given in the Chapter Problem (n = 106, $\bar{x}$ = 98.20, s= 0.62) and a 0.05 significance level, test the claim that the mean body temperature of healthy adults is equal to 98.6° F.

$H_0$: $\mu$ = 98.6 (Claim) $H_1$: $\mu \neq 98.6$

1. Press **STAT<TESTS>1:Z-Test** for a screen similar to screen (1). Make sure that the input line (Inpt:) has Stats highlighted by using ▶ and pressing **ENTER**. Use ▼ to input $\mu_0$:98.6, and since n = 106 > 30, use ▼ and let $\sigma$ = s = 0.62. Also input $\bar{x}$ and **n** as in screen (1), highlight the the alternate hypothesis $\mu:\neq \mu_0$, and then press **ENTER**.

2. Highlight 'Calculate' and press **ENTER** for screen (2). If you had highlighted Draw in the last line of screen (1) and then pressed **ENTER**, you would have obtained screen (3). Both screens (2) and (3) reveal that the test statistic z = ⁻6.6423. In screen (3) this value is far to the left of the given standard normal distribution since -3.5 is about the minimum value shown on the z axis.

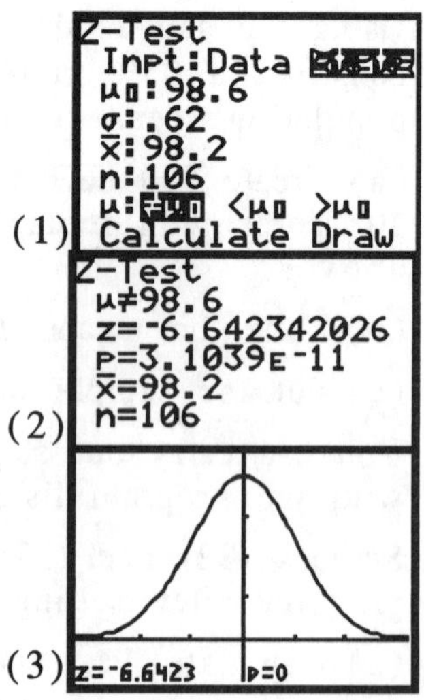

3. The Home screen calculation could be done using
   $z = (\bar{x} - \mu_0)/(\sigma/\sqrt{(n)}) = (98.2 - 98.6)/(0.62/\sqrt{(106)})$
   $= {}^{-}6.6423$ as in the top lines of screen (4).
4. The critical z value can be found in Table A-2 of the text or calculated as in the last few lines of screen (4) using invNorm (with **↑DISTR 3** as in Chapter 5) for $z_{\alpha/2} = \pm 1.96$.

   Since ${}^{-}6.6423 < {}^{-}1.96$ we reject the null hypothesis and conclude that μ is statistically significantly less than 98.6.

```
(98.2-98.6)/(0.6
2/√(106))
            -6.642342026
invNorm(0.025)
            -1.959963986
invNorm(1-0.025
             1.959963986
```
(4)

## P-value Method [pg. 387]

**EXAMPLE 1b** [pg. 390]: Use the p-value method to test the claim of the preceding example.

Following steps 1 and 2 above, you obtained screen (2), which gives a p-value = 3.1039E-11 = 0.000000000031039, given in screen (3) as p=0 to four decimal places. Since the p-value < 0.05, we reject the null hypothesis, as before.

**Note:** When the p-value is part of the output, there is no need to calculate or look up the critical values. This makes it a more efficient approach than the traditional method.

**Note:** The area shaded in each tail of the distribution in screen (3) is half of 3.1039E-11, but this is too small to show on the screen. Screen(5) gives another example of a two-tail test, in which the area is split between both tails, for the above data if $H_0$: μ = 98.3. In this case we would fail to reject the null hypothesis since the p-value = 0.0968 >0.05.

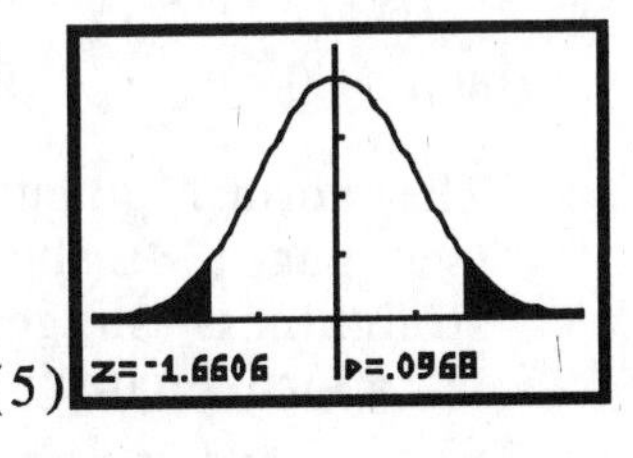

(5)

## One-Tail Test

**EXAMPLE Coke Volumes** [pg. 385]: Data Set 1 [in Appendix B] includes the volumes (in ounces) of the Coke in a sample of 36 different cans that are all labeled 12 oz. Here are the sample statistics: n = 36, $\bar{x}$ = 12.19 oz, s = 0.11 oz. Using a 0.01 significance level, test the manager's claim that the mean is greater than 12 oz.

$H_0$: μ ≤ 12    $H_1$: μ > 12 (Claim)

1. Press **STAT<TESTS>1:Z-Test** for a screen similar to screen (6). Make sure that the input line (Inpt:) has Stats highlighted by using ▶ and pressing **ENTER**. Use ▼ to input $\mu_0$:12, and since n = 36 > 30, use ▼ and let σ = s = .11.

   Also input $\bar{x}$ and **n** as in screen (6), highlight the Alternate Hypothesis μ:> $\mu_0$, and press **ENTER**.
2. Highlight Calculate and press **ENTER** for screen (7), with test statistic of z = 10.36 and p-value = 1.85E-25.

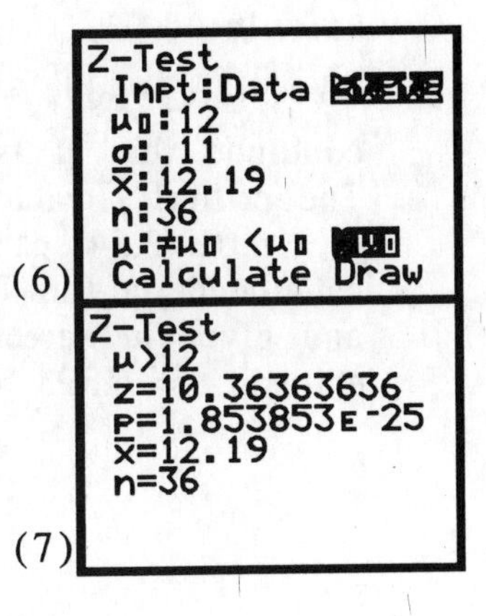

(6) (7)

3. The critical z value is 2.33, as shown in screen (8) using ↑**DISTR 3** as in Chapter 5 and 6. Since 10.36 > 2.33, we reject $H_0$ and conclude that the sample data supports the claim that the mean amount of Coke is greater than 12 oz. The very small p-value (much less than 0.01) supports the same conclusion.

```
invNorm(1-.01)
     2.326347877
```
(8)

## TESTING A CLAIM ABOUT A MEAN: SMALL SAMPLES [pg. 399]

**EXAMPLE Pulse Rates** [pg. 402]: A teacher [the Author], at the peak of an exercise program, claimed that his pulse rate was lower than the mean pulse rate of statistics students. The teacher's pulse rate was measured to be 60 beats in one minute, and the 20 students in his class measured their pulse rates with the results listed below. (Store these scores in L1.) At the 0.01 level of significance, test the claim that this sample comes from a population with a mean that is greater than 60 beats in one minute.

L1. = { 61 74 72 77 79 90 69 79 90 72 68 64 83 82 83 62 88 74 55 65 }

$H_0$: $\mu \leq 60$ $H_1$: $\mu > 60$ (Claim)

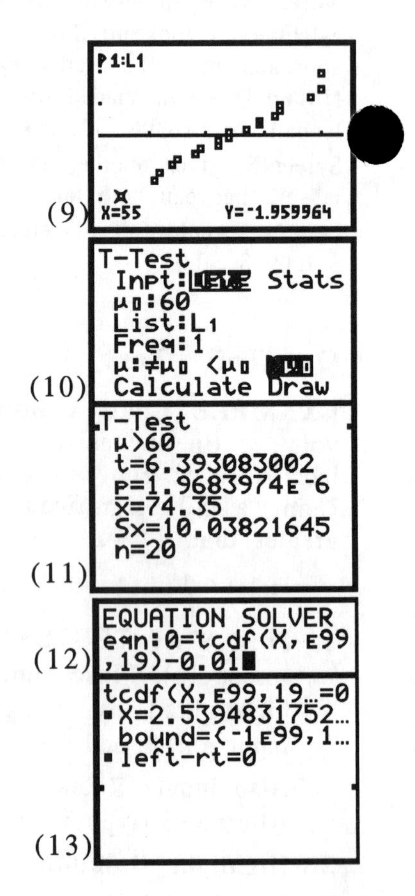


1. The normal quantile plot of the data in screen (9) (see page 38) shows no extreme outliers and its straightness suggest it comes from a population that is approximately normally distributed.

2. Press **STAT**<**TESTS**>**2:T-Test** for a screen similar to screen (10). Make sure that the input line (Inpt:) has Data highlighted by using ◄ and pressing **ENTER**. Use ▼ to input $\mu_0$:60, paste L1 for the data List, and set Freq to 1. Highlight the alternate hypothesis $\mu$:> $\mu_0$, and press **ENTER**.

3. Highlight Calculate and press **ENTER** for screen (11), with the test statistic of t = 6.393 and the p-value = 0.00000197. The sample statistics $\bar{x}$ and Sx = s are also calculated.

4. Since the p-value = 0.0000776 <0.01, we reject $H_0$ and conclude that $\mu$ is significantly greater than 60 bpm. The critical t value with 19 degrees of freedom and $\alpha$ = 0.01 (2.54) can be found in Table A-3 of the text or calculated on the TI-83 Plus as explained on page 40 and given in screens (12) and (13). Since the test statistic t = 6.393 > 2.54, it also supports the claim.

# TESTING A CLAIM ABOUT A PROPORTION [pg. 409]

**EXAMPLE Survey of Voters** [pg. 411]: In a survey of 1002 people, 701 said that they voted in the recent presidential election. Test the claim that when surveyed, the proportion of people who say that they voted is equal to 0.61, which is the proportion of people who actual did vote. Use the 0.05 significance level.

$H_0: p = 0.61$ (Claim) $H_1: p \neq 0.61$

1. With nq=1002*0.39 =391 >5 a z-distribution can be used.
2. Press **STAT**<TESTS>**5**:1-PropZTest for a screen similar to screen (14). Complete the information as shown.
3. Highlight Calculate and press **ENTER** for screen (15), with the test statistic z = 5.815 and a p-value = 6.08E-9. $\hat{p} = x / n = 0.6996 = 0.70$.
4. Home screen calculations are as follows:
   $z = (\hat{p} - p_0)/(\sqrt{(p_0 * (1 - p_0)/n)}$
   $= (0..6996 - 0.61)/\sqrt{(0.61*0.39/1002)} = 5.815$
5. Since the p-value = 0.000000006 < 0.05, we reject $H_0$ and conclude that a significant larger proportion of people say they vote than actually do. The critical values of ± 1.96 (as in screen (4)) lead to the same conclusion with 5.815 > 1.96.

```
1-PropZTest
 p0:.61
 x:701
 n:1002
 prop≠p0 <p0 >p0
 Calculate Draw
```
(14)

```
1-PropZTest
 prop≠.61
 z=5.814983935
 p=6.0830259E-9
 p̂=.6996007984
 n=1002
```
(15)

# TESTING A CLAIM ABOUT A STANDARD DEVIATION OR A VARIANCE [pg. 417]

**EXAMPLE IQ Scores of Statistic Professors** [pg. 419]: For a simple random sample of adults, IQ scores are normally distributed with a mean of 100 and a standard deviation of 15. A simple random sample of 13 statistic professors yield a standard deviation of s = 7.2. Use a 0.05 significance level to test the claim that $\sigma = 15$. $H_0: \sigma = 15$ $H_1: \sigma \neq 15$

1. Calculate the test statistic from the following equation.
   $\chi^2 = (n - 1) s^2/\sigma^2 = (13 - 1)7.2^2/15^2 = 2.7648 < 4.404 = \chi^2_L$
2. Calculate the p-value as in screen (18) with $\chi^2$cdf from **↑DISTR 7** for p = 0.006. This is a two-tail test.
3. Since the p-value = 0.006 < 0.05, and we are in the left critical region (with $\chi^2_L$ = 4.404 as calculated in screens (16) and (17)), we reject the null hypothesis. The group 'statistics professors' has significant less variation than the general population.

```
EQUATION SOLVER
eqn:0=χ²cdf(0,X,
12)-0.025
```
(16)

```
χ²cdf(0,X,12)…=0
•X=4.4037885081…
 bound={-1E99,1…
•left-rt=1E-14
```
(17)

```
(13-1)7.2²/15²
            2.7648
χ²cdf(0,2.7648,1
2)
       .0030109923
Ans*2
       .0060219847
```
(18)

# 8 Inferences from Two Samples

In Chapter 6 and 7 you found confidence intervals and tested hypotheses for data sets that involved only one sample from one population. In this chapter you will learn how to calculate test statistics and confidence intervals for data sets that involve two samples from two populations. As in Chapters 6 and 7 you will use **STAT<TESTS>** functions. To complete the problems in this Chapter successfully, you need to be familiar with Data list and sample Statistics input, and Calculate and Draw output possibilities, from the examples in Chapters 6 and 7.

## INFERENCES ABOUT TWO MEANS: INDEPENDENT AND LARGE SAMPLES [pg. 438]

**EXAMPLE Coke versus Pepsi** [pg. 440]: Use a 0.01 significance level to test the claim that the mean weight of regular Coke is different than the mean weight of regular Pepsi. The original data are listed in Data Set 1 [Appendix B], and the summary statistics are listed at the right.

| Regular Coke | Regular Pepsi |
|---|---|
| n1 = 36 | n2 = 36 |
| $\bar{x}1$ = 0.81682 | $\bar{x}2$ = 0.82410 |
| s1 = 0.007507 | s2 = 0.005701 |

$H_0$: $\mu_1 = \mu_2$ (or $\mu_1 - \mu_2 = 0$) $H_1$: $\mu_1 \neq \mu_2$

1. Press **STAT<TESTS>3**:2-SampZTest and set up the resulting screen as in screen (1), using $\sigma 1$ = s1 and $\sigma 2$ = s2 because of the large sample sizes.
2. Highlight Calculate and press **ENTER** for screen (2), with z = ⁻4.6338 and the p-value = 0.00000359.
3. The Home screen calculation is as follows.

   $z = ((\bar{x}1 - \bar{x}2) - (\mu_1 - \mu_2))/\sqrt{(\sigma 1^2/n1 + \sigma 2^2/n2)}$

   $= (0.81682 - 0.82410)/\sqrt{(0.007507^2/36 + 0.005701^2/36)}$
   
   $= {}^{-}4.6338$
4. Since the p-value = 0.00000359 is less than the significance level 0.01, we would reject the null hypothesis. The mean weight of the regular Coke is (statistically) significantly smaller than the mean weight of the Pepsi.

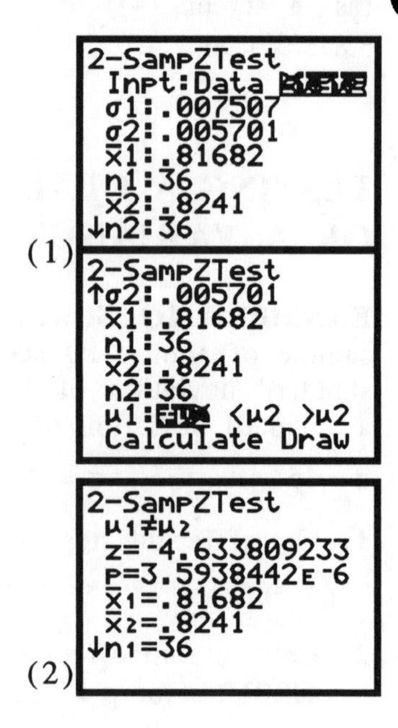


### Confidence Interval [pg. 441]

**EXAMPLE Coke and Pepsi** [pg. 442]: Using the sample data given in the preceding example, construct a 99% confidence interval estimate of the difference between the mean weights of regular Coke and regular Pepsi.

1. Press **STAT**<TESTS>**9**:2-SampZInt and set up the resulting screen as in screen (3) using $\sigma 1 = s1$ and $\sigma 2 = s2$ because of the large sample sizes.
2. Highlight Calculate and press **ENTER** for screen (4), with the 99% confidence interval equal to (-.0113, -.0032), or $-.0113 < \mu_1 - \mu_2 < -0032$.

   You can use $\bar{x}1 - \bar{x}2 = .81682 - .82410 = -.0073$ to find the margin of error E by subtracting it from the larger interval value. $E = -.0032 - (-.0073) = .004$, so you can also write the 99% confidence interval as $-.0073 \pm .004$.
3. The Home screen calculation uses $z_{\alpha/2} = 2.576$ for

   $E = z_{\alpha/2} * \sqrt{(\sigma 1^2/n1 + \sigma 2^2/n2)}$

   $= 2.576 * \sqrt{(.007507^2/36 + .005701^2/36)} = .004$ as above.

```
2-SampZInt
 Inpt:Data Stats
 σ1:.007507
 σ2:.005701
 x̄1:.81682
 n1:36
 x̄2:.8241
↓n2:36
```
(3)
```
2-SampZInt
↑σ2:.005701
 x̄1:.81682
 n1:36
 x̄2:.8241
 n2:36
 C-Level:.99
 Calculate
```
```
2-SampZInt
 (-.0113,-.0032)
 x̄1=.81682
 x̄2=.8241
 n1=36
 n2=36
```
(4)

## INFERENCES ABOUT TWO MEANS: MATCHED PAIRS [pg. 448]

**EXAMPLE Do Male Students Exaggerate Their Heights?** [pg. 451]: Using the sample data below [Table 8-1] with the outlier (third pair) deleted, test the claim that male statistics students do exaggerate by reporting heights that are greater that their actual measured heights (in inches). Use a 0.05 significance level.

| Subject | A | B | C | D | E | F | G | H | I | J | K | L |
|---|---|---|---|---|---|---|---|---|---|---|---|---|
| Reported height L1 | 68 | 74 | x | 66.5 | 69 | 68 | 71 | 70 | 70 | 67 | 68 | 70 |
| Measured height L2 | 66.8 | 73.9 | x | 66.1 | 67.2 | 67.9 | 69.4 | 69.9 | 68.6 | 67.9 | 67.6 | 68.8 |
| Difference L3=L1-L2 | 1.2 | 0.1 | x | 0.4 | 1.8 | 0.1 | 1.6 | 0.1 | 1.4 | -0.9 | 0.4 | 1.2 |

$H_0$: $\mu_d \le 0$ $H_1$: $\mu_d > 0$ Claim

Put the reported heights in L1 and the measured heights in L2. Store the difference of L1 - L2 in L3 (i.e., highlight L3 in the **STAT 1**:Editor, type L1 – L2, and then press **ENTER** or on the home screen, type L1 – L2 **STO▸**L3 then press **ENTER**) to obtain the results in the last line of the table above.

1. Press **STAT**<TESTS>**2**:T-Test and set up the resulting screen as in screen (5).
2. Highlight Calculate and press **ENTER** for screen (6), with t = 2.701 and the p-value = 0.0111.
   Note that $\bar{d} = \bar{x} = 0.6727$ and $s_d = Sx = 0.8259$, which could also be calculated with **STAT[CALC]1**:1-Var Stats L3.
3. The Home screen calculation is as follows:

   $t = (\bar{d} - \mu)/(s/\sqrt{(n)}) = (0.6727 - 0)/(0.8259/\sqrt{(11)}) = 2.701$

```
T-Test
 Inpt:Data Stats
 μ0:0
 List:L3
 Freq:1
 μ:≠μ0 <μ0 >μ0
 Calculate Draw
```
(5)
```
T-Test
 μ>0
 t=2.701377679
 p=.0111303292
 x̄=.6727272727
 Sx=.8259429872
 n=11
```
(6)

**4.** Since p-value = 0.0111 is less than the significance level 0.05, we would reject the null hypothesis. There is sufficient evidence to support the claim that male statistics students exaggerated their heights by reporting values that are greater than their measured heights.

## Confidence Intervals for $\mu_d$ [pg. 452]

**EXAMPLE** [pg. 453]: Use the sample data from the preceding example to construct a 95% confidence interval estimate of $\mu_d$.

1. Press **STAT**<TESTS>**8**:TInterval and set up the resulting screen as in screen (7).
2. Highlight Calculate and press **ENTER** for screen (8), with the 95% confidence interval equal to (0.11785,1.2276), or $0.11785 < \mu_d < 1.2276$.

   The margin of error E can be calculated by taking the difference between $\bar{d}$ and the larger interval value. E = 1.2276 – 0.6727 = 0.5549, so we can also express the 95% confidence interval as 0.6727 ± 0.5549.
3. The Home screen calculation uses $t_{\alpha/2}$ = 2.228 (with 11 – 1 = 10 degrees of freedom), so

   $E = t_{\alpha/2} * s/\sqrt{(n)} = 2.228*0.8259/\sqrt{(11)} = 0.5548$ as above.

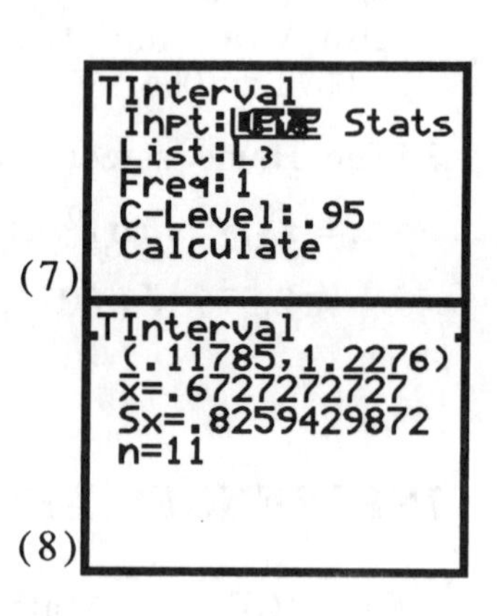

(7)
(8)

# INFERENCES ABOUT TWO PROPORTIONS [pg. 458]

**EXAMPLE Viagra Treatment and Placebo** [pg. 460]: In preliminary tests for adverse reactions, it was found that when 734 men were treated with Viagra, 16% of them experienced headaches. Among 725 men in a placebo group, 4% experienced headaches. Is there sufficient evidence to support the claim that among those men who take Viagra, headaches occur at a rate that is greater than the rate for those who do not take Viagra? Use a 0.01 significance level.

Viagra group: $\hat{p}_1 = 0.16$  n1 = 734  x1= 0.16 *734 = 117

Placebo group: $\hat{p}_2 = 0.04$  n2 = 725  x2 = 0.04*725 = 29

$H_0$: $p_1 \le p_2$ (or $p_1 - p_2 \le 0$)  $H_1$: $p_1 > p_2$ Claim

1. Press **STAT**<TESTS>**6**:2-PropZTest and set up the resulting screen as in screen (9).
2. Highlight Calculate and press **ENTER** for screen (10), with z = 7.599 and p-value = 0.000000000000015.
3. The Home screen calculation is as follows:

   p = (x1 + x2)/(n1 + n2) = (117 + 29)/(734 + 725) = 0.1001

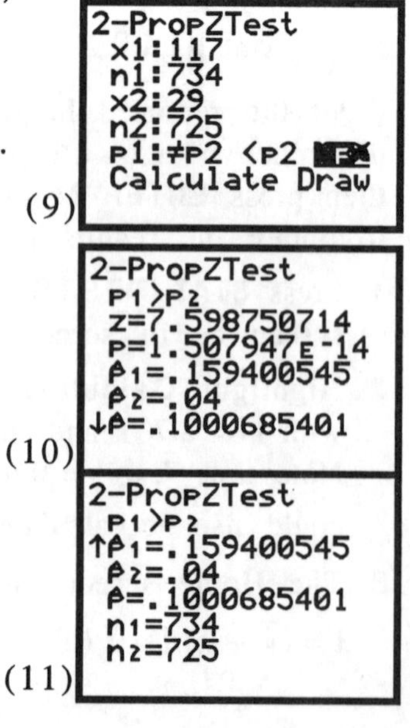

(9)
(10)
(11)

$z = ((\hat{p}_1 - \hat{p}_2) - (p_1 - p_2))/\sqrt{(p*q/n1 + p*q/n2)}$
$= (0.16 - 0.04)/\sqrt{(0.10*0.90/734 + 0.10*0.90/725)}$
$= 7.6$

**4.** Since the p-value = 0.000+ is less than the significance level 0.01, we would reject the null hypothesis. The headache rate is significantly greater for men treated with Viagra than for the men who took a placebo.

### Confidence Interval [pg. 463]

**EXAMPLE** [pg. 463]: Use the data in the preceding example to construct a 99% confidence interval for the difference between the two population proportions.

1. Press **STAT<TESTS>B**:2-PropZInt and set up the resulting screen as in screen (12).
2. Highlight Calculate and press **ENTER** for screen (13), with the 99% confidence interval equal to (0.07987, 0.15893), or $0.07987 < p_1 - p_2 < 0.15893$. You can use $\hat{p}_1 - \hat{p}_2 = 0.16 - 04 = 0.12$ to find the margin of error E by subtracting it from the larger interval value. E = 0.1589 − 0.12 = 0.039, so you can also write the 99% confidence interval as 0.12 ± 0.04.

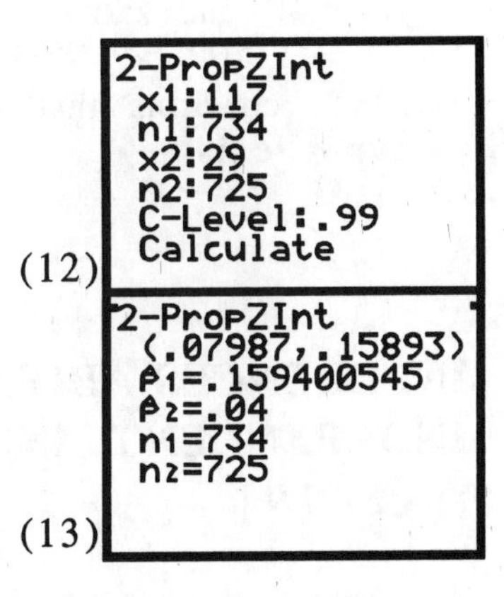

(12)
(13)

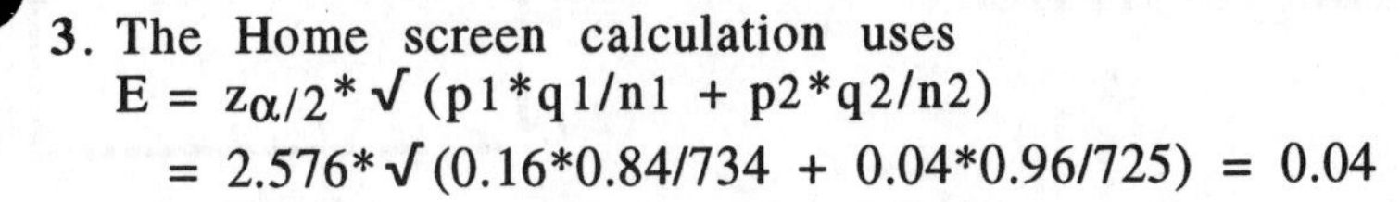

3. The Home screen calculation uses
   $E = z_{\alpha/2}*\sqrt{(p1*q1/n1 + p2*q2/n2)}$
   $= 2.576*\sqrt{(0.16*0.84/734 + 0.04*0.96/725)} = 0.04$

## COMPARING VARIATION IN TWO SAMPLES [pg. 470]

**EXAMPLE Coke versus Pepsi** [pg. 473]: Data Set 1 [in Appendix B] includes the weights (in pounds) of samples of regular Coke and regular Pepsi. The sample statistics are summarized in the accompanying table. Use a 0.05 significance level to test the claim that the weights of regular Coke and the weights of regular Pepsi have the same standard deviation.

| Regular Coke | Regular Pepsi |
|---|---|
| n1 = 36 | n2 = 36 |
| $\bar{x}$1 = 0.81682 | $\bar{x}$2 = 0.82410 |
| s1 = 0.007507 | s2 = 0.005701 |

$H_0$: $\sigma_1 = \sigma_2$ (or $\sigma_1/\sigma_2 = 1$)  $H_1$: $\sigma_1 \neq \sigma_1$

1. Press **STAT<TESTS>D**:2-SampFTest and set up the resulting screen as in screen (14).

2-SampFTest
Inpt:Data Stats
Sx1:.007507
n1:36
Sx2:.005701
n2:36
σ1:≠σ2 <σ2 >σ2
Calculate Draw

(14)

2. Highlight Calculate and press **ENTER** for screen (15), with F = 1.7339 and the p-value = 0.1082 which is also part of the output of the Draw option in screen (16).

2-SampFTest
σ1≠σ2
F=1.733926762
p=.1081582463
Sx1=.007507
Sx2=.005701
↓n1=36

(15)

**Note**: If you reverse the order in which you enter the data, as in screen (17), for the results in screen (18), with F = 0.5767 (in the left tail of screen (16)) but the same p-value = 0.1082. (Also note that 1/0.5767256275 = 1.733926762.)

3. The Home screen calculation is as follows:
   $F = s1^2/s2^2 = 0.007507^2/0.005701^2 = 1.7339$ with $(n1 - 1, n2 - 1) = (35, 35)$ degrees of freedom.

4. Since the p-value = 0.1082 is greater than the significance level 0.05, we would fail to reject the null hypothesis. The standard deviations are not (statistically) significantly different.

   The critical value for the right tail can be calculated with the **MATH 0**:Solver... similar to the example on page 40 and shown in screens (19) and (20). Since $F_R = 1.9611 > 1.7339$ we are not in the critical region which confirms that we would fail to reject the null hypothesis.

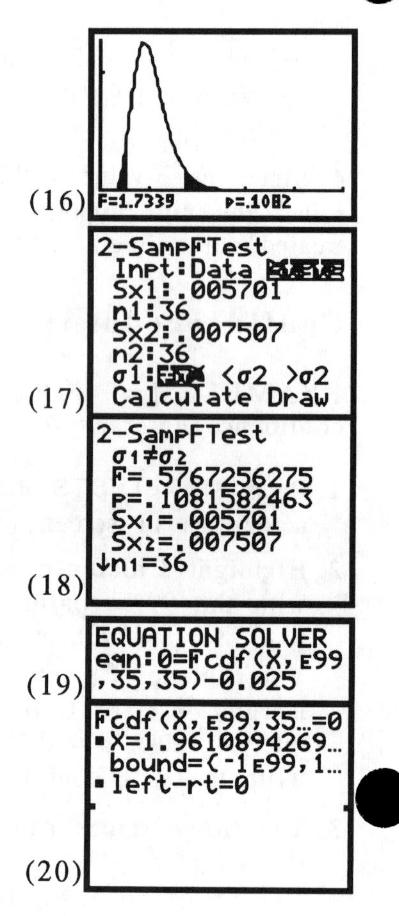

(16) (17) (18) (19) (20)

## INFERENCES ABOUT TWO MEANS: INDEPENDENT AND SMALL SAMPLES [pg. 479]

**Assumptions**

1. The two samples are independent.
2. The two samples are randomly selected from normally distributed populations.
3. At least one of the two samples is small ($n \leq 30$).

### Case 1: Both Population Variances Are Known [pg. 479]

Because of the assumptions above, this will be a z-test, even with the small sample size, and will be treated exactly as the large-sample case described on page 48.

### Case 2: Equal Variances (We Fail to Reject $\sigma_1^2 = \sigma_2^2$) [pg. 481]

If the variances are equal, you can pool (combine) the sample variances to get a weighted average of $s_1^2$ and $s_2^2$, which is the best possible estimate of the variance $\sigma^2$ that is common to both populations.

**EXAMPLE Cigarette Filters and Nicotine**: [pg. 481]: Use the sample statistics that follow at the 0.05 level of significance to test the claim that the mean amount of nicotine in filtered king-size cigarettes is equal to the mean amount of nicotine for nonfiltered king-size cigarettes. (All measurements are in milligrams.)

Filtered kings: $n_1 = 21$, $\bar{x}_1 = 0.94$, $s_1 = 0.31$

Nonfiltered kings: $n_2 = 8$, $\bar{x}_2 = 1.65$, $s_2 = 0.16$

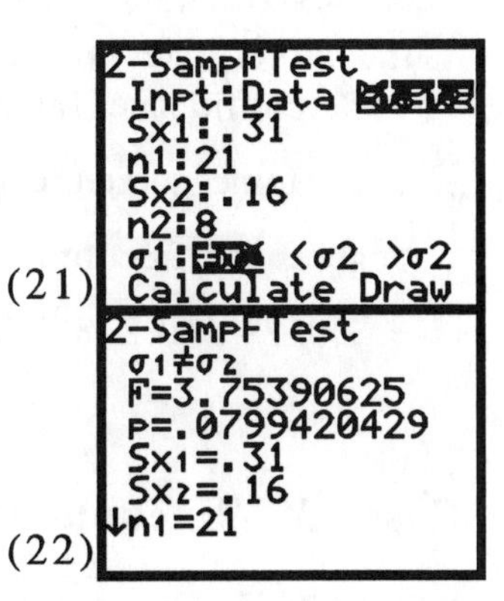


(21)

(22)

## Preliminary F Test ($H_0$: $\sigma_1 = \sigma_2$)

1. Press **STAT**<TESTS>**D**:2-SampFTest and set up the resulting screen as in screen (21).
2. Highlight Calculate and press **ENTER** for screen (22), with the p-value = 0.0799.
3. Since 0.0799 > 0.05, we fail to reject $H_0$: $\sigma_1 = \sigma_2$. This is Case 2, where we pool the variance.

## Test $H_0$: $\mu_1 = \mu_2$ (or $\mu_1 - \mu_2 = 0$) $H_1$: $\mu_1 \neq \mu_2$

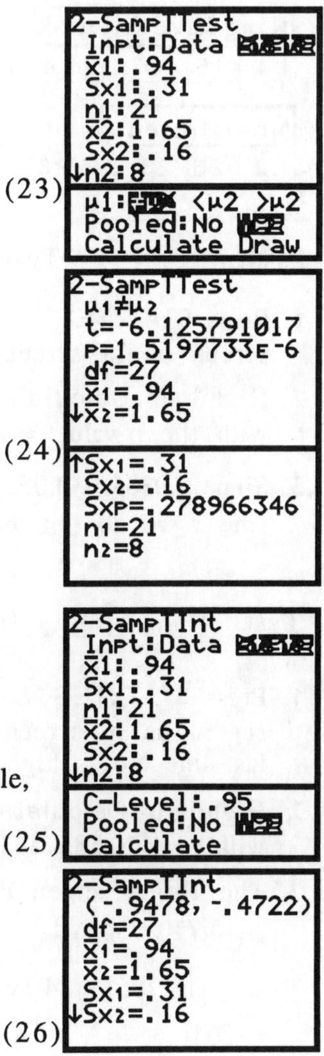


(23)

(24)

(25)

(26)

1. Press **STAT**<TESTS>**4**:2-SampTTest and set up the resulting screen as in screen (23). Notice that in the second-to-last line, Yes is highlighted for Pooled.
2. Highlight Calculate and press **ENTER** for screen (24), with t = ⁻6.126, p-value = 0.0000015198, and $s_p$ = Sxp = 0.27897 (for $s_p^2 = 0.0778$).
3. The Home screen calculation is as follows:

$$s_p^2 = ((n_1-1)s_1^2 + (n_2-1)s_2^2)/((n_1-1) + (n_2-1))$$
$$= (20*0.31^2 + 7*0.16^2)/(20 + 7) = 0.0778$$
$$t = ((\bar{x}1 - \bar{x}2) - (\mu_1 - \mu_2))/\sqrt{(s_p^2/n1 + s_p^2/n2)}$$
$$= (0.94 - 1.65)/\sqrt{(0.27897^2/21 + 0.27897^2/8)}$$
$$= {}^-6.127$$

4. Since the p-value = 0.0000015198 is less than the significance level 0.05, we would reject the null hypothesis. The filtered cigarettes have significantly less nicotine than the nonfiltered ones.

## Confidence Interval

**EXAMPLE** [pg. 483]: Using the data in the preceding example, construct a 95% confidence interval estimate of $\mu_1 - \mu_2$.

1. Press **STAT**<TESTS>**0**:2-SampTInt and set up the resulting screen as in screen (25).
2. Highlight Calculate and press **ENTER** for screen (26), with the 95% confidence interval equal to (⁻0.9478, ⁻0.4722) or $^-0.9478 < \mu_1 - \mu_2 < {}^-0.4722$.

$\bar{x}1 - \bar{x}2 = 0.94 - 1.65 = {}^-0.71$ to find the margin of error E by subtracting it from the larger interval value. $E = {}^-0.4722 - ({}^-0.71) = 0.2378$, so you can also write the 95% confidence interval as $^-0.71 \pm 0.2378$.

3. The Home screen calculation uses $t_{\alpha/2} = 2.052$ (with $n1 + n2 - 2 = 21 + 8 - 2 = 27$ degrees of freedom) for $E = t_{\alpha/2} * \sqrt{(sp^2/n1 + sp^2/n2)}$
$= 2.052 * \sqrt{(0.27897^2/21 + 0.27897^2/8)} = 0.2378$ as above.

## Case 3: Unequal Variances (Reject $\sigma_1^2 = \sigma_2^2$) [pg. 484]

**EXAMPLE Cigarette Filters and Tar** [pg. 485]: Use the following sample data at the 0.05 level of significance to test the claim that the mean amount of tar in filtered king-size cigarettes is less than the mean amount of tar for nonfiltered king-size cigarettes. (All measurements are in milligrams.)

King-size filtered

| L1 | 16 | 15 | 16 | 14 | 16 | 1 | 16 | 18 | 10 | 14 | 12 | 11 | 14 | 13 | 13 | 13 | 16 | 16 | 8 | 16 | 11 |
|---|---|---|---|---|---|---|---|---|---|---|---|---|---|---|---|---|---|---|---|---|---|

Nonfiltered

| L2 | 23 | 23 | 24 | 26 | 25 | 26 | 21 | 24 |
|---|---|---|---|---|---|---|---|---|

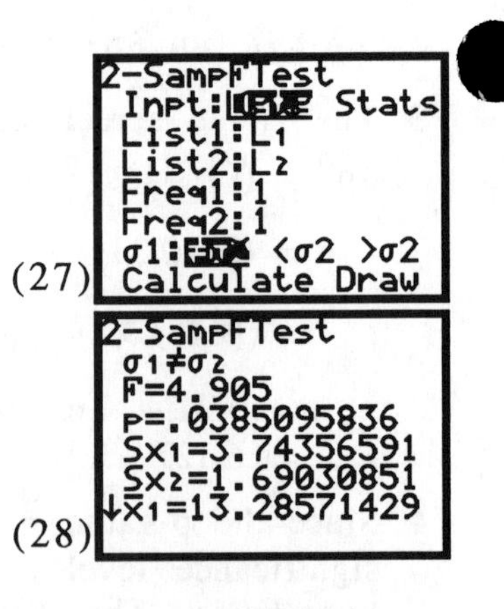

(27)

(28)

### Preliminary F Test ($H_0$: $\sigma_1 = \sigma_2$)

1. Press **STAT**<TESTS>**D**:2-SampFTest and set up the resulting screen as in screen (27).
2. Highlight Calculate and press **ENTER** for screen (28), with the p-value = 0.0385.
3. Since 0.0385 < 0.05, we reject Ho: $\sigma 1 = \sigma 2$. This is Case 2, where we do <u>not pool</u> the variance.

### Test $H_0$: $\mu_1 \geq \mu_2$ (or $\mu_1 - \mu_2 \geq 0$) $H_1$: $\mu_1 < \mu_2$

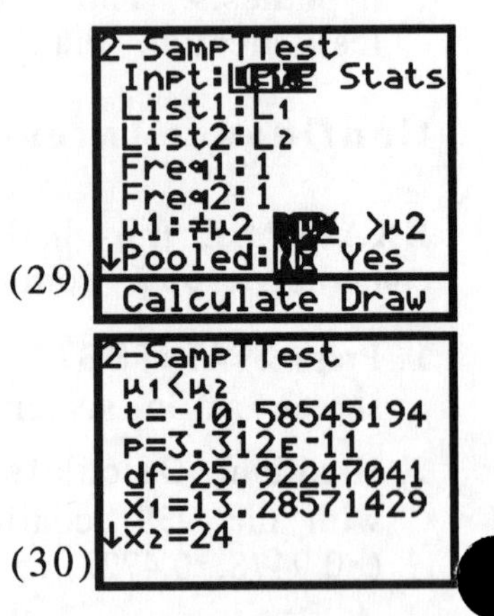

(29)

(30)

1. Press **STAT**<TESTS>**4**:2-SampTTest and set up the resulting screen as in screen (29). Notice that in the second-to-last line No is highlighted for Pooled.
2. Highlight Calculate and press **ENTER** for screen (30), with $t = {}^-10.585$ and the p-value = 0.0000000003312.
3. The Home screen calculation is as follows:

$t = ((\bar{x}1 - \bar{x}2) - (\mu_1 - \mu_2))/\sqrt{(s1^2/n1 + s2^2/n2)}$

$= (13.285 - 24)/\sqrt{(3.74^2/21 + 1.69^2/8)} = {}^-10.59$

**4.** Since the p-value = $0.000^{+}$ is less than the significance level 0.05, we would reject the null hypothesis. The filtered cigarettes have significantly less tar than the nonfiltered ones.
**Note**: Since the raw data is used here, the TI-83 Plus carries more decimals than the calculations in the text.

## Confidence Interval

**EXAMPLE** [pg. 486]: Use the data in the preceding example to construct a 95% confidence interval estimate of $\mu_1 - \mu_2$.

1. Press **STAT<TESTS>0**:2-SampTInt and set up the resulting screen as in screen (31).
2. Highlight Calculate and press **ENTER** for screen (32), with the 95% confidence interval equal to $({}^{-}12.8, {}^{-}8.633)$, or ${}^{-}12.8 < \mu_1 - \mu_2 < {}^{-}8.633$.
   You can use $\bar{x}1 - \bar{x}2 = 13.2857 - 24 = {}^{-}10.714$ to find the margin of error E by subtracting it from the larger interval value. $E = {}^{-}8.633 - ({}^{-}10.714) = 2.081$, so you can also write the 95% confidence interval as ${}^{-}10.714 \pm 2.081$.

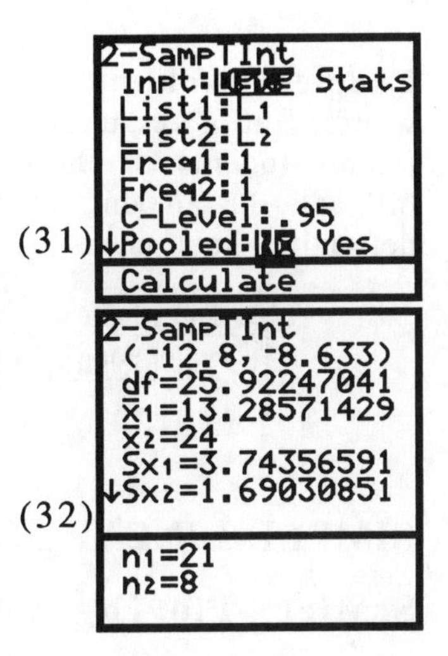

(31)
(32)

3. The Home screen calculation uses

$$E = t_{\alpha/2} * \sqrt{(s1^2/n1 + s2^2/n2)}$$
$$= 2.060 * \sqrt{(3.74^2/21 + 1.69^2/8)} = 2.08 \text{ as above.}$$

$t_{\alpha/2} = 2.060$ is from df = 25 (conservative estimate from 25.922 of screen (32)). You can use Formula 8-1 of the text [pg. 484] to calculate degrees of freedom (df):

$$df = (A+B)^2/[A^2/(n_1 - 1) + B^2/(n_2 - 1)]$$

where $A = s_1{}^2/n_1$ and $B = s_2{}^2/n_2$.
See the calculation in screen (33).

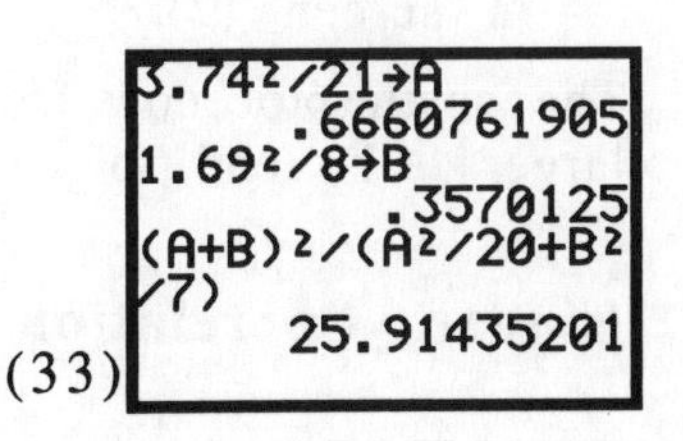

(33)

**Note**: The text used df = 8 − 1 = 7 with a critical t of 2.391 for the conservative (wider) confidence interval of $({}^{-}13.1, {}^{-}8.3)$

# 9 Correlation and Regression

Our study of correlation and regression will be aided by the **STAT<TESTS>E:** LinRegTTest function. This function is used for linear regression with Y as a straight-line function of X. Program **A2MULREG** applies to multiple regression where Y is a function of X1, X2, . . ., Xn and supplements the computer output given in the text.

**Chapter Problem Tipping** [pg. 505]: How much do people tip a waiter or waitress at a restaurant? The table below gives a sample of six bills and tips paid [from students of the author]. Is there a relationship between the amount of the bill and the amount of the tip? If there is a relationship, how do we use it to determine how much of a tip should be left?

| x: Bill (dollars) | L1 | 33.46 | 50.68 | 87.92 | 98.84 | 63.60 | 107.34 |
|---|---|---|---|---|---|---|---|
| y: Tip (dollars) | L2 | 5.50 | 5.00 | 8.08 | 17.00 | 12.00 | 16.00 |

## SIMPLE LINEAR REGRESSION AND CORRELATION

### Scatter Plot [pg. 507]

1. Enter data into L1 and L2.
2. Set up Plot1 for a scatter plot as in screen (1).
3. Press **ZOOM 9**:ZoomStat and **TRACE** for the scatter plot in screen (2).

The scatterplot does seem to reveal a pattern showing that larger bills tend to go along with larger tips.

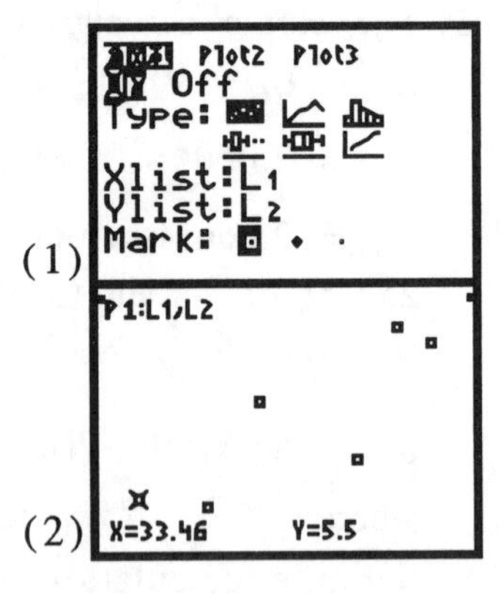

(1)

(2)

### Linear Correlation Coefficient r [pg. 509]

**EXAMPLE Tipping** [pg. 510]: Using the previous data, find the value of the linear correlation coefficient r.

1. Press **STAT<TESTS>E**:LinRegTTest and set up the resulting screen as in screen (3). (Y1 is pasted from **VARS<YVARS>1**: Function **1**: Y1.)
2. Highlight Calculate in the last line, and press **ENTER** for the two screens of output (4) and (5) with r = 0.8282 in the last line of screen (5).

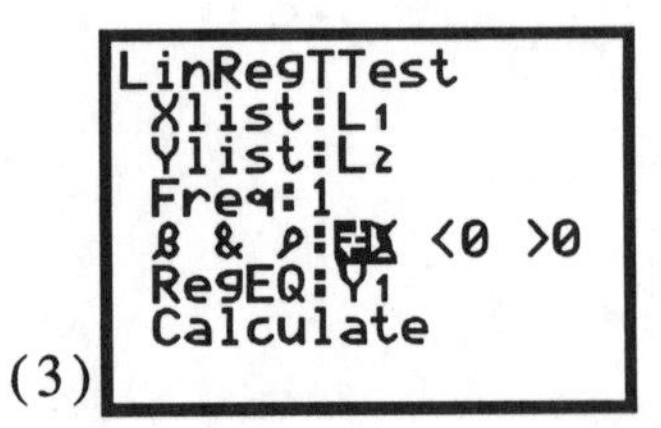

(3)

**Note**: There are other ways of finding r on the TI-83 Plus, but this way will be the most productive in the long run.

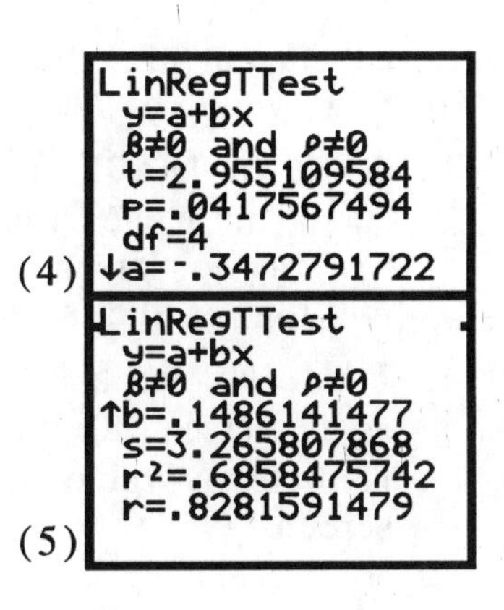


## Explained Variation $r^2$ (Coefficient of Determination) [pgs. 512, 540]

**EXAMPLE Tipping** [pg. 512]: Referring to bill/tip data, what proportion of the variation in the tip can be explained by the variation in the bill?

From screen (5), second last line we see the $r^2 = .6858$ so about 69% of the variation in tips can be explained by the linear relationship between the bill and the tip.

## Formal Hypothesis Test of the Significance of r [pg. 514]

**EXAMPLE Tipping** [pg. 515]: Using the data above, test the claim that there is significant linear correlation between the amount of the bill and the amount of the tips.

$H_0: \rho = 0 \quad H_1: \rho \neq 0$

Using the first screen of the output of **STAT<TESTS>E**: LinRegTTest (screen (4)), we see the test statistic t = 2.955 and the p-value = 0.0418 < 0.05, indicating significant positive linear correlation.

## Regression [pg. 525]

**EXAMPLE Tipping** [pg. 527]: For the chapter problem, you used the given data (x = amount of bill and y = amount of tip) to find that the linear correlation coefficient of r = 0.8282 indicates that there is significance linear correlation. Now find the regression equation of the straight line that relates x and y.

Using the two screens of output, (4) and (5) above with $a = b_0 = {}^-0.3473$ and $b = b_1 = 0.1486$,

thus $\hat{y} = -0.3473 + 0.1486x$.

Set up Plot1 as in screen (1). The regression equation (RegEQ) is automatically stored and turned on as $Y_1$ in the **Y =** editor by step 1 of screen (3). All other plots must be turned off. After step 2 of screen (3) press **ZOOM 9:ZoomStat** and then **TRACE** for the least square regression line plotted with the scatter plot of points as in screen (6).

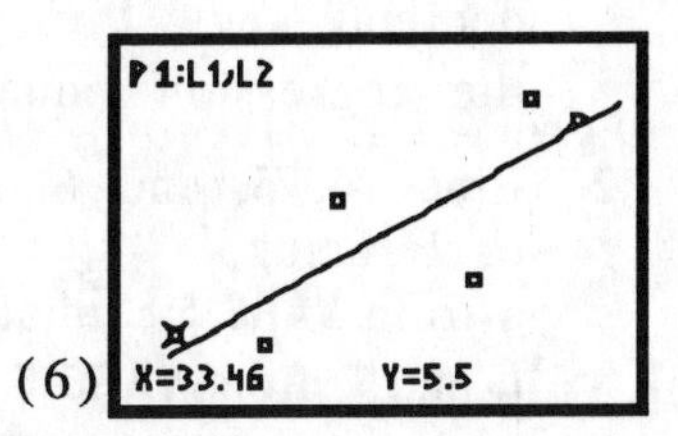


## Predictions [pg. 528]

**EXAMPLE Tipping** [pg. 529]: Using the sample data of the chapter problem, you found that there is significant linear correlation between amount of bills and amount of tips, and you also found the regression equation as above.

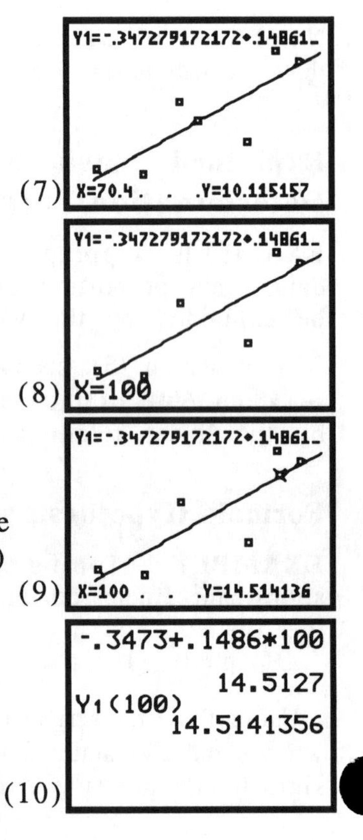


If the amount of a bill is \$100, predict the amount of the tip that would be left.

1. While in **TRACE** mode as in screen (6), press ▼ for the cursor to move from a data point to the regression line, as in screen (7). (The equation listing starts at the top of the screen.)
2. Type **100**, and a large X = 100 shows in the bottom line as in screen (8). Press **ENTER** for screen (9), with a predicted y value of \$14.51, as in the bottom line of screen (9).

You can also calculate the predictive value on the Home screen by typing in the regression equation as in the top lines of screen (10) or by pasting $Y_1$ to the Home screen and then typing **(100)** and pressing **ENTER**, as in the last line of screen (10). ($Y_1$ comes from **VARS**<Y-VARS>**1**: Function **1**: $Y_1$)

**Note**:List RESID is automatically created by running LinRegTTest for those who wish to look at the residuals.

## Prediction Intervals for y [pg. 543]

**EXAMPLE Tipping** [pg. 543]: The chapter problem in previous sections, you have seen that there is significant linear correlation and that when x = \$100, the predicted y value is \$14.52. Construct a 95% prediction interval for the tip, given that the bill is \$100. From Table A-3 of the text, $t_{\alpha/2} = 2.776$ using 6 − 2 degrees of freedom. (Do the following after **STAT**<TESTS>**E**: LinRegTTest of screen (3)).

1. The first lines of screen (11) show the bill (\$100) stored in X, 2.776 stored in T, and the predictive value (14.51 in this example) stored in Y, with $Y_1$ (X) **STO▸**Y. These setup values are tied together with colons (above the decimal key). Pressing **ENTER** reveals Y = \$14.51 from the regression equation $Y_1$.
2. Lines 4, 5, and 6 of screen (11) show the calculation of E using:
   s from **VARS 5**:Statistics<TEST>**0**:s
   n and $\bar{x}$ from **VARS 5**:Statistics<XY>
   $\Sigma x^2$ and $\Sigma x$ from **VARS 5**:Statistics<Σ>.
   Pressing **ENTER** reveals E = \$10.46.
3. Screen (12) reveals the 95% prediction interval of \$4.05 < y < \$24.98. That range is relatively large. One factor contributing to the large range is the small sample size of 6.

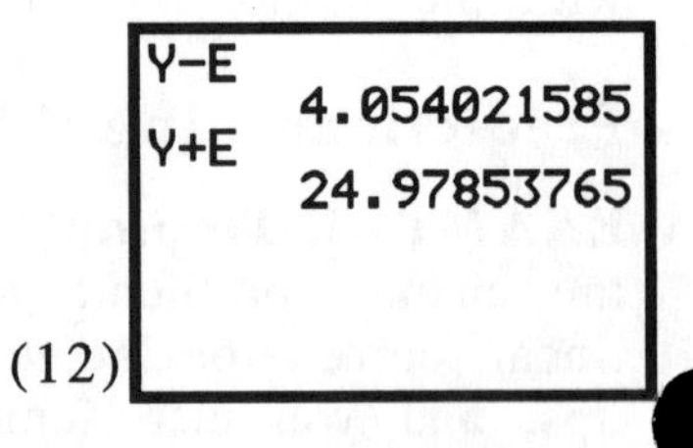

**Note**: The predictive interval for different bills (X) and levels (T) are easily calculated using the last-entry feature. First, recall the top input lines of screen (11) so you can make changes. Then press **ENTER** and recall the lines that calculate E in screen (11).

**Note**: Calculating predictive intervals for simple linear regression and for multiple regression is automated in the program of the next section.

## MULTIPLE REGRESSION [pg. 549]

**EXAMPLE Bears** [pg. 550] Measurements taken from bears [Table 9-3] are given below. Find the multiple regression equation in which the dependent (y) variable is weight and the independent variables are head length (HEADL) and total overall length (LENGTH).

| Y or C1 | C2 | C3 | C4 | C5 | C6 | C7 |
|---|---|---|---|---|---|---|
| WEIGHT | AGE | HEADL | HEADW | NECK | LENGTH | CHEST |
| 80 | 19 | 11 | 5.5 | 16 | 53 | 26 |
| 344 | 55 | 16.5 | 9 | 28 | 67.5 | 45 |
| 416 | 81 | 15.5 | 8 | 31 | 72 | 54 |
| 348 | 115 | 17 | 10 | 31.5 | 72 | 49 |
| 262 | 56 | 15 | 7.5 | 26.5 | 73.5 | 41 |
| 360 | 51 | 13.5 | 8 | 27 | 68.5 | 49 |
| 332 | 68 | 16 | 9 | 29 | 73 | 44 |
| 34 | 8 | 9 | 4.5 | 13 | 37 | 19 |

**Note**: The TI-83 Plus calculator did not have multiple regression features when this companion was written. There is such a feature for the TI-89 which has a STAT Editor similar to the TI-83 Plus so a supplement might be available as an application that can be stored [pg. 552]. The TI-89 version has easy to follow menu input screens.

Program A2MULREG is introduced below, and its output is compared to the Minitab output. But first we must store the data into the matrix [D]. (The availability of program A2MULREG is given in the Appendix.) One important point to remember about data input is that the dependent variable Y, WEIGHT in this example, must be in the first column of [D]. Minitab has no such restriction.

### Enter Data into Matrix [D]

The data can be entered from a computer or from another TI-83 Plus (see the Appendix). The following method is for entering the data from the keyboard.

Press **MATRX** [EDIT] **4**:[D] and then **8 ENTER 7 ENTER** for screen (13). You might have to use the **DEL** key when inputting the size of the matrix (8 rows by 7 columns), depending on the size of the previous matrix. Although you may have different values in your matrix, leave them as is because you are going to enter values over them.

With the cursor in the first cell of the first row (as shown in screen 13), enter the data row by row by typing 80 **ENTER** 19 **ENTER** 11 **ENTER** 5.5 **ENTER** 16 **ENTER** 53 **ENTER** 26 **ENTER** for the first row of the data table. With the last **ENTER**, the cursor will return to the first slot in the second row. Enter the other seven rows in the same manner to the last value, as shown in screen (14).

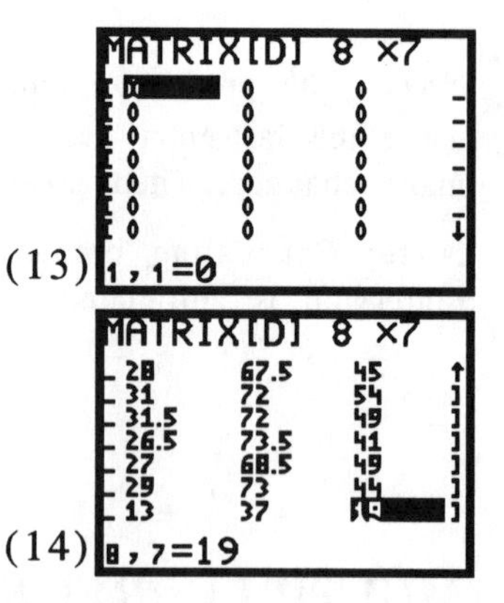

(13)
(14)

**EXAMPLE Minitab Output** [pg. 551]: Verify the Minitab output with program A2MULREG (for a model with the two independent variables HEADL in C3 and LENGTH in C6) for the following important components:

1. The multiple regression equation WEIGHT = -374 + 18.8 HEADL + 5.87 LENGTH
2. Adjusted $R^2 = 0.759$
3. The overall significance of the multiple regression equation or p = 0.012

**Program A2MULREG**

1. Press **PRGM**<NAMES> and use ▼ to highlight A2MULREG. Press **ENTER**, and prgmA2MULREG is pasted to the Home screen. Press **ENTER** again for screen (16), which reminds you how to enter data in matrix [D] and alerts you to the fact that data in other matrices would be lost.
2. Press **ENTER** for screen (17) which gives you three options. Press **1** for MULT REGRESSION.
   **Note**: Press **3** to quit if your data is not in matrix [D].
3. The next screen (18) requires the number of independent variables (2) and the column numbers in [D] in which they are located (3 and 6).
4. Pressing **ENTER** after the 6 above gives screen (19), with the overall significance of the multiple regression equation: **P = 0.012**, $R^2$ = 0.828, and $R^2$ (adjusted), or **R-SQ(ADJ) = 0.7592.**
5. Press **ENTER** for screen (20). Use the B0 and COEFFs to come up with the regression equation:
   **WEIGHT = -374.3 + 18.820 HEADL + 5.875 LENGTH**
6. Press **ENTER** for screen (21), which gives the option to

(15)
```
prgmA2MULREG
```
(16)
```
DATA IN MAT [D]
COL Y,X1,X2,..XN
Y MUST BE IN THE
1ST COL OF [D].
[A],[B],[C],[D],
[E]+[F] USED.
```
(17)
```
MULT REG+CORR
1:MULT REGRESSIO
2:CORR MATRIX
3:QUIT
```
(18)
```
HOW MANY IND VAR
?2
COL. OF VAR.
               1
?3
COL. OF VAR.
               2
?6
```
(19)
```
  DF SS
RG 2  113142.268
ER 5  23505.7315
     F=12.03
     P=.012
  R-SQ=.828
  (ADJ).7592
S=68.56490584
```
(20)
```
B0=-374.3034756
CL COEFF / T  P
3 18.82040176
   .81      .453
6 5.874757415
   1.16     .299
```

QUIT; then do quit.

**Note:** The MAIN MENU of screen (21) gives the options of doing confidence and predictive intervals and of doing residual analysis.

(21)

## Correlation Matrix

**EXAMPLE** [pg. 553]: What is the best regression equation to use (the greatest $R^2$) if only a single independent variable is to be used?

```
MULT REG+CORR
1:MULT REGRESSIO
2:CORR MATRIX
3:QUIT
```
(22)

```
[[1    .814 .88...
 [.814 1    .87...
 [.884 .879 1  ...
 [.885 .896 .96...
 [.971 .906 .95...
 [.897 .802 .91...
 [.992 .839 .87...
```
(23)

1. When you run program A2MULREG, one of the first screens (screens (17) and (22)) gives the option of calculating the correlation matrix.

   **Note**: Pressing **ENTER** from the Home screen will restart the program for screen (16) if the last thing done on the Home screen was to quit this program.

2. Press **2**:CORR MATRIX, but be aware of the busy calculating indicator in the upper-right corner of the screen—these calculations take awhile. The partial output is given in screen (23). (You can view the rest of the output by using the cursor control keys.) The first column is duplicated in the table with headings at the right. CHEST is the variable that has the greatest linear correlation with WEIGHT and thus the greatest $r^2$, with r = 0.992 ($r^2$ = .983).

| | WEIGHT |
|---|---|
| WEIGHT | 1 |
| AGE | 0.814 |
| HEADL | 0.884 |
| HEADW | 0.885 |
| NECK | 0.971 |
| LENGTH | 0.897 |
| CHEST | 0.992 |

# 10 Multinomial Experiments and Contingency Tables

In this chapter you will use the STAT Editor or spreadsheet to calculate the $\chi 2$ statistic for multinomial experiments and the **STAT<TESTS>C**:$\chi^2$-Test function to do contingency table analysis.

## MULTINOMIAL EXPERIMENTS: GOODNESS-OF-FIT [pg. 575]

**EXAMPLE M&Ms:** [pg. 581]: Mars, Inc., claims that its M&M candies are distributed with the color percentages of 30% brown, 20% yellow, 20% red, 10% orange, 10% green, and 10% blue. The colors of the M&Ms listed in Data Set 10 [of Appendix B] are summarized in the table below. Using the sample data and a 0.05 significance level, test Mars's claim about the color distribution of M&M candies.

| | Brown | Yellow | Red | Orange | Green | Blue |
|---|---|---|---|---|---|---|
| Observed frequency (L1) | 33 | 26 | 21 | 8 | 7 | 5 |
| Expected proportion (L3) | 0.3 | 0.2 | 0.2 | 0.1 | 0.1 | 0.1 |

### To Calculate Chi-square Statistic $\Sigma[(O-E)^2/E]$

1. Put the observed frequencies in L1 and the expected proportions in L3.
2. Highlight L2 as at the top of screen (1), and type **L3*** sum(**L1** as at the bottom of the screen (sum under **↑LIST<MATH>5**).
3. Press **ENTER** for the expected frequencies (E) in L2, as in screen (2). (Of course, if these values are given or if they are easy to calculate without the spreadsheet, just enter them into L2.)
4. Highlight L4 and type (L1−L2)²/L2 (as in the last line of screen (2)), and then press **ENTER** for the chi-square contribution of each color in L4, as in screen (3). The last row (color blue) makes the largest contribution, 2.5.
5. Press **↑QUIT** to return to the Home screen.

   Sum the above contributions with sum(**L4**, and then press **ENTER** for the chi-square statistic, which is 5.95 (see screen (4)).

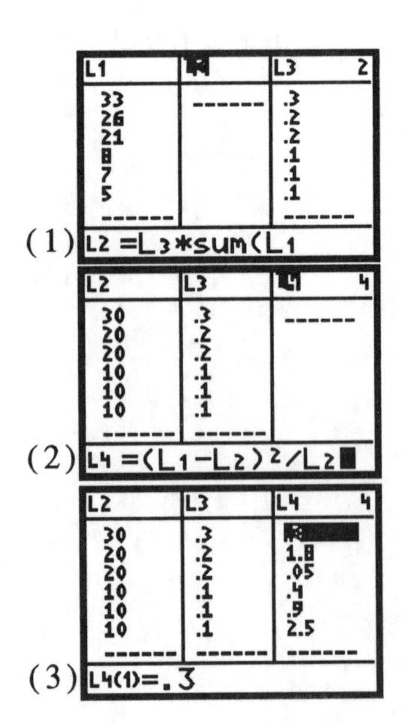

## P-Value [pg. 584]

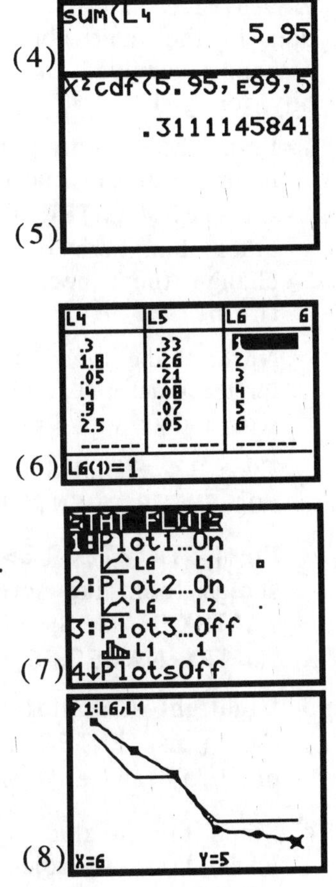


6. Press **↑DISTR** 7:χ2cdf(**5.95,E99,5** and then **ENTER** for a p-value = 0.311 df = 5, or 6 − 1 (see screen (5)).
7. Since the p-value = 0.311 is larger than the significance level α = 0.05, we would fail to reject Mars's claim.

To graphically compare the observed frequencies with the expected frequencies, put the integers 1 to 6 in L6 as in screen (6).

**Note**: The observed proportions could be calculated in L5 by highlighting L5, typing L1/sum(L1), then pressing **ENTER**. (The first value in L5, or .33, compares with the the first expected proportion in L3 (screen (3) or .30).

Set up Plot1 and Plot2 as in screen (7) with different marks (a square for observed values and a dot for expected). Press **ZOOM 9:ZoomStat**, and then press **TRACE** and ▶ a few times for screen (8).

Notice that the points for each color are relatively close together. The last color (blue) has the greatest discrepancy, but this was found not significant above.

**Note**: The observed proportions in L5 and the expected proportions in L3 could have been plotted (see Figure 10.6 of text [pg. 583]), but the graph would look the same—only the y-axis scale would change.

## CONTINGENCY TABLES [pg. 589]

**TABLE [10-1] Titanic Sinking** [pg. 573]: The table below summarizes the fate of the passengers and crew when the Titanic sank on April 15, 1912. "Let's stipulate that the data are sample data randomly selected from the population of all theoretical people who would find themselves in the same conditions... for the purposes of this discussion and analysis [pg. 573].

| | Men | Women | Boys | Girls | Total |
|---|---|---|---|---|---|
| Survived | 332 (20%) | 318 (75%) | 29 (45%) | 27 (60%) | 706 (32%) |
| Died | 1360 (80%) | 104 (25%) | 35 (55%) | 18 (40%) | 1517 (68%) |

**EXAMPLE Titanic Expected Frequency** [pg. 592]: Find the expected frequencies for the lower-left cell, assuming independence between the row variable (whether the person survived) and the column variable (whether the person is a man, woman, boy, or girl).

E =(row total)*(column total)/(grand total)
=(1360 + 104 + 35 + 18)*(332 + 1360)/(332 + 318 + 29 + 27 + 1360 + 104 + 35 + 18) = 1154.64

**Note**: This method is not efficient, but the output of **STAT<TESTS> C:χ2-Test** below gives all of the expected values automatically (see screen (14)).

**EXAMPLE Titanic Sinking** [pg. 593]: At the 0.05 significance level, use the data in the table above to test the claim that when the Titanic sank, whether someone survived or died is independent of whether the person is a man, woman, boy, or girl.

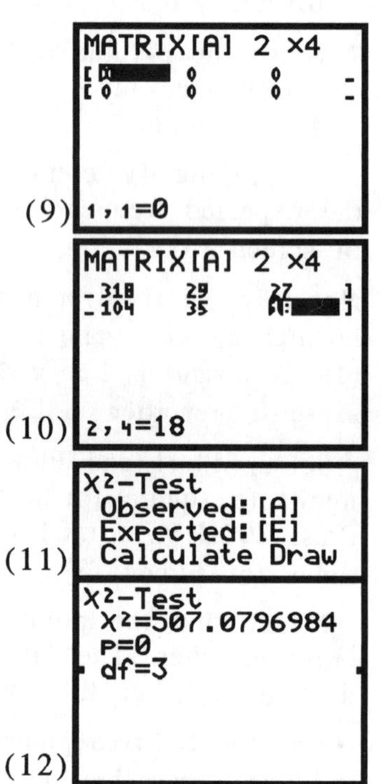


1. Enter the data in a matrix. (Use matrix [A] as the Data matrix, but any matrix will do.) Press **↑MATRX<EDIT> 1**:[A] **2 ENTER 4 ENTER** for screen (9). You might have values other than zero in your matrix, but don't bother to change them because you will be entering values over them.

   **Note**: If the size of the previous matrix had more than one digit for the number of rows or columns, you will have to delete the second digit with **DEL**.

   Type 332 **ENTER** 318 **ENTER** 29 **ENTER** 27 **ENTER** 1360 **ENTER** 104 **ENTER** 35 **ENTER** 18 **ENTER** for screen (10).

2. Press **STAT<TESTS> C:$\chi^2$–Test** and have the resulting screen look like screen (11) with [A] pasted by pressing **↑MATRX**<NAMES> **1**:[A] and [E] pasted with **↑MATRX**<NAMES> **5**:[E].

3. Highlight Calculate in the last line of screen (11), and then press **ENTER** for screen (12) with a test statistic of 507.08 and a p-value = 0.

4. Since the p-value = $0.000^+$ is less than the significance level 0.05, we reject the null hypothesis. It appears that whether a person survived the Titanic and whether that person is a man, women, boy, or girl are dependent variables.

To compare the observed and expected frequencies on the same screen, press **MODE** and make the resulting screen look like screen (13) by highlighting **0** decimals in the second row and pressing **ENTER**. Press **↑QUIT** to return to the Home screen.

**Note**: Do not forget to change the **MODE** back to Float decimal when you have finished the following.

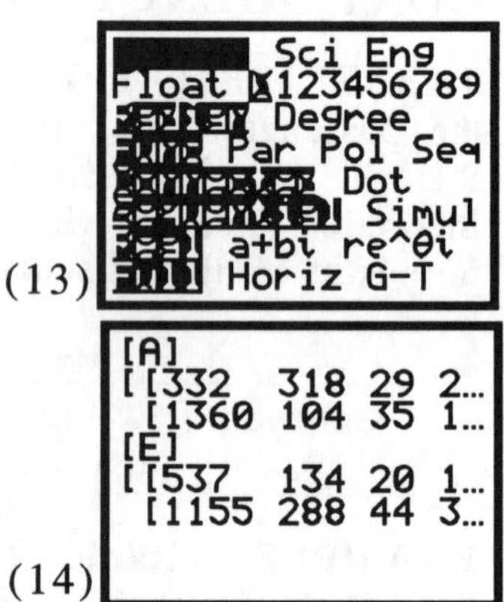


Paste [A] to the Home screen, and press **ENTER**. Do the same thing for matrix [E] for screen (14).

Notice that 332 men survived, less than the 537 expected by the null hypothesis. More women survived, 318, than the expected 134. Or looking at the original sample, 20% of the men and 75% of the women survived. These are significantly different from each other and from the 32% for each group that is expected under the null hypothesis. There did seem to be some success in saving the women and children first.

# 11 Analysis of Variance

In this chapter you will learn about the **STAT<TEST>F:ANOVA** function for doing one-way ANOVA problems. You will also be introduced to the program A1ANOVA, which will extend your ability to do two-way ANOVA for two-factor designs (with an equal number of observations in each cell). The purpose is to duplicate the Minitab ANOVA tables in the text with the TI-83 Plus. Follow the text for the proper interpretation of these tables.

## ONE-WAY ANOVA [pg. 617]

The following procedure works for both equal and unequal sample sizes.

**EXAMPLE Head Injuries in Car Crashes** [pg. 618]: Given the sample data below [Table 11-1] and a significance level of a = 0.05, use the TI-83 Plus calculator to test the claim that the four samples come from populations with the same mean ($H_0$: $\mu_1 = \mu_2 = \mu_3 = \mu_4$).

| | | | | | |
|---|---|---|---|---|---|
| Subcompact (L1) | 681 | 428 | 917 | 898 | 420 |
| Compact (L2) | 643 | 655 | 442 | 514 | 525 |
| Midsize (L3) | 469 | 727 | 525 | 454 | 259 |
| Full Size (L4) | 384 | 656 | 602 | 687 | 360 |

1. Put the data in L1 to L4 as indicated in the table.
2. Press **STAT<TESTS>F:ANOVA** for ANOVA( on the Home screen. Type L1,L2,L3,L4 for screen (1).
3. Press **ENTER** for screens (2) and (3), with the test statistic F = 0.9922 and a p-value = 0.4216 in screen (2) and the pooled standard deviation, or Sxp = 172.36, in screen (3). The results will be similar to those in the Minitab and other displays in the text [pg. 619].

(1)
```
ANOVA(L1,L2,L3,L
4
```

(2)
```
One-way ANOVA
 F=.9921670139
 p=.4215699156
 Factor
  df=3
  SS=88425
↓ MS=29475
```

(3)
```
One-way ANOVA
↑ MS=29475
 Error
  df=16
  SS=475323.2
  MS=29707.7
 Sxp=172.359218
```

**Note**: Program A1ANOVA, which is introduced in the next section, gives the means and standard deviations of the raw data stored in the lists. The program accepts sample summary statistics (means, standard deviations, and sample sizes) as an input option in addition to the raw data option. The TI-89 (see note on the bottom of page 59) also accepts summary statistics and has the capability of doing two-way ANOVA.

## TWO-WAY ANOVA [pg. 630]

The following example uses program A1ANOVA for two-way ANOVA. The availability of program A1ANOVA is given in the Appendix.

## Two-Factor Design with an Equal Number of Observations per Cell

**EXERCISES 4 TO 6 Length Estimates by Gender and Major** [pgs. 638, 639]: The sample data below are student estimates (in feet) of the length of their classroom. The actual length of the classroom is 24 ft 7.5 in.

| Estimates of Lengths (in feet) of Classroom Categorized by Gender and Major | | | |
|---|---|---|---|
| | B (1) Math | B (2) Business | B (3) Liberal Arts |
| Female A (1) | 28 | 35 | 40 |
| | 25 | 25 | 21 |
| | 30 | 20 | 30 |
| Male A (2) | 25 | 30 | 25 |
| | 30 | 24 | 20 |
| | 20 | 25 | 32 |

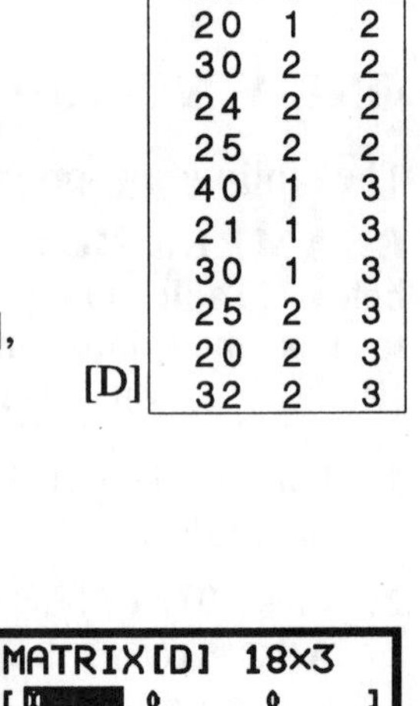

[D]

| | | |
|---|---|---|
| 28 | 1 | 1 |
| 25 | 1 | 1 |
| 30 | 1 | 1 |
| 25 | 2 | 1 |
| 30 | 2 | 1 |
| 20 | 2 | 1 |
| 35 | 1 | 2 |
| 25 | 1 | 2 |
| 20 | 1 | 2 |
| 30 | 2 | 2 |
| 24 | 2 | 2 |
| 25 | 2 | 2 |
| 40 | 1 | 3 |
| 21 | 1 | 3 |
| 30 | 1 | 3 |
| 25 | 2 | 3 |
| 20 | 2 | 3 |
| 32 | 2 | 3 |

The data in the table above is going to be stored in matrix [D], with 18 rows and 3 columns as shown at the right. All the data is in the first column with the first six being estimates by Math majors (indicated by six 1s in the third column). These are followed by the data for Business and then Liberal Arts majors. The second column indicates gender with 1 = female, 2 = male.

**1. Enter Data into Matrix [D].**

(a) Press **↑MATRX**<EDIT>**4**:[D] for the Edit screen of matrix [D]. Type **18 ENTER 3 ENTER** for 18 rows and 3 columns, and the cursor will be at the first slot of the matrix as in screen (4). If your screen does not have all zeros, don't worry, you will be typing over these values.

(b) You can enter data by row by typing **28** and then pressing **ENTER** followed by **1 ENTER 1 ENTER**, which brings us to the second row as in screen (5). Continue with the 17 other rows given above with **25 ENTER 1 ENTER 1 ENTER** ...**32 ENTER 2 ENTER 3 ENTER**.

**2. Running Program A1ANOVA.**

(a) Press **PRGM**<EXEC> and highlight program A1ANOVA. Press **ENTER**, and prgmA1ANOVA is pasted to the Home screen as in screen (6).

(b) Press **ENTER** for option screen (7). Press **3**:2WAY FACTORIAL for a Two-Way Factor ANOVA design and the instructional screen (8), which reminds you how to store the data in matrix [D].

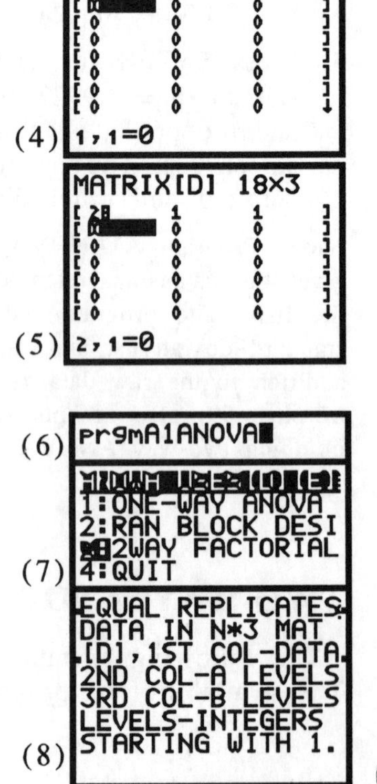

(c) Press **ENTER** for screen (9), which gives you two options: continue, or quit if the data is not ready.

(d) Press **1**, since the data is in [D], for the ANOVA table of screen (10).
**Note**: A busy signal shows in the upper-right-hand corner of the display screen while the calculations are being done.

Screen (10) has for gender: F(A) = 0.78, and p-value = 0.395. Major: F(B) = 0.13 and with p-value = 0.876 ( Press **ENTER** for screen (11)). Interaction F(AB) = 0.19 with a p-value of 0.832 also in screen (11). These all agree with the Minitab display in the text [pg. 639].

**Note**: Although screen (10) does not give Mean Squares (MS), they are easily calculated by dividing Sum of Squares (SS) by Degrees of Freedom (DF).

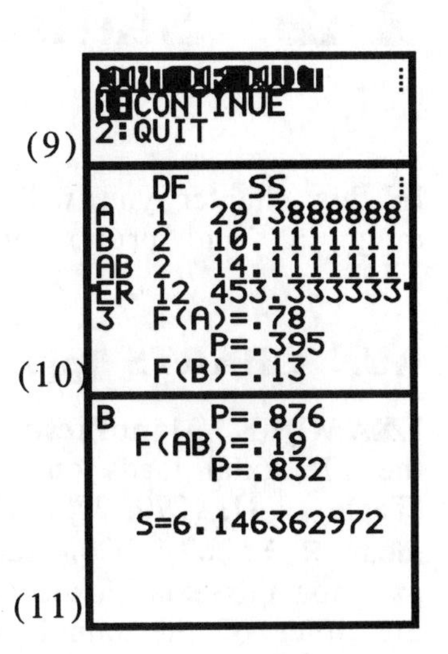


**Special Case: One Observation per Cell and No Interaction [pg. 636]**

**EXERCISES 9 TO 10 Beams Manufactured by 4 Operators on 3 Machines** [pg. 640]: The sample data below are the number of support beams manufactured by four different operators using each of three different machines. Assume that there is no interaction effect from operator and machine. Match the Minitab display.

**Number of Beams Manufactured Categorized by Operator and Machine**

| A | Machine 1 | Machine 2 | Machine 3 |
|---|---|---|---|
| Operator 1 | 66 | 74 | 67 |
| Operator 2 | 58 | 67 | 68 |
| Operator 3 | 65 | 71 | 65 |
| Operator 4 | 60 | 64 | 66 |
| mean | 25.5 | 27 | 30.25 |
| S.D. | 5.97 | 6.27 | 7.68 |

[D]

| | | |
|---|---|---|
| 66 | 1 | 1 |
| 58 | 2 | 1 |
| 65 | 3 | 1 |
| 60 | 4 | 1 |
| 74 | 1 | 2 |
| 67 | 2 | 2 |
| 71 | 3 | 2 |
| 64 | 4 | 2 |
| 67 | 1 | 3 |
| 68 | 2 | 3 |
| 65 | 3 | 3 |
| 66 | 4 | 3 |

The contents of a 12 x 3 matrix [D] are given with all the data in the first column, followed by the operator number in the second column and the machine number in the third column.

Running program A1ANOVA 3:2WAY FACTORIAL gives screens (12) and (13) of output, which are similar to the Minitab display in the text [pg. 640].

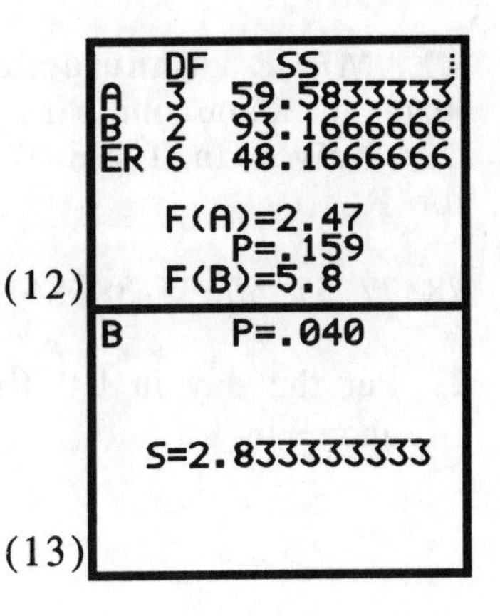

# 12 Statistical Process Control

In this chapter you will learn how to plot run charts and control charts for range, $\bar{x}$, and proportions.

## RUN CHARTS [pg. 655]

**EXAMPLE Manufacturing Cola Cans** [pg. 655]: Store the 175 axial loads on aluminum cans in row order from [Table 12-1], {270, 273,...274, 292}, into list LALCAN with mean $\bar{x}$ = 267.1. Construct a run chart by using a vertical axis for the axial loads and a horizontal axis to identify the order of the sample data.

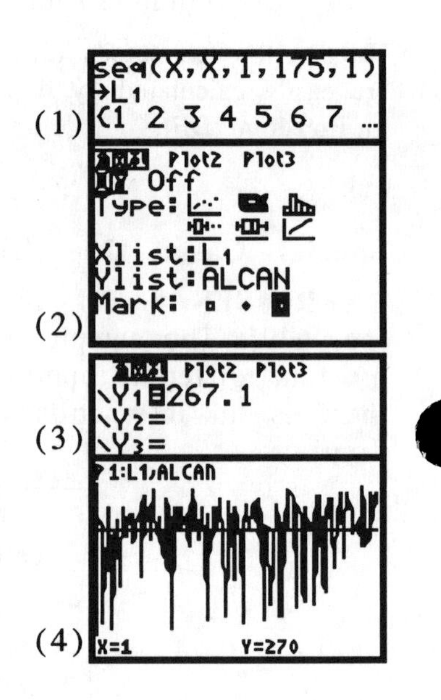


1. Generate integers from 1 to 175 and store them in L1 with **↑LIST**<OPS>5:seq(X,X,1,175,1) **STO▸** L1. Then press **ENTER** (see screen (1)).
2. Set up Plot1 as an xy-Line plot as in screen (2). (All other Stat Plots should be off.) Paste **ALCAN** as on page 8 (step 6).
3. Press the **Y=** key and let Y1= 267.1 (the mean axial load) as in screen (3). (All other Y = plots should be off.)
4. Press **ZOOM 9**:ZoomStat, and adjust the **WINDOW** to better fill the screen by changing Xmin = 0 and Xmax = 175. Then press **TRACE** for screen (4), which is similar to the plot in the text [pg. 655]).

## CONTROL CHART FOR MONITORING VARIATION: THE R CHART [pg. 658]

**EXAMPLE Manufacturing Cola Cans** [pg. 659]: Refer to the ranges of the axial loads of aluminum cans (of samples of size 7 collected each day for 25 working days) given in Table 12-1 of the text and repeated below. Construct a control chart for R.

78 77 31 50 33 38 84 21 38 77 26 78 78 17 83 66 72 79 61 74 64 51 26 41 31

1. Put the day in L1 (integers from 1 to 25) and the corresponding ranges from above in L2.

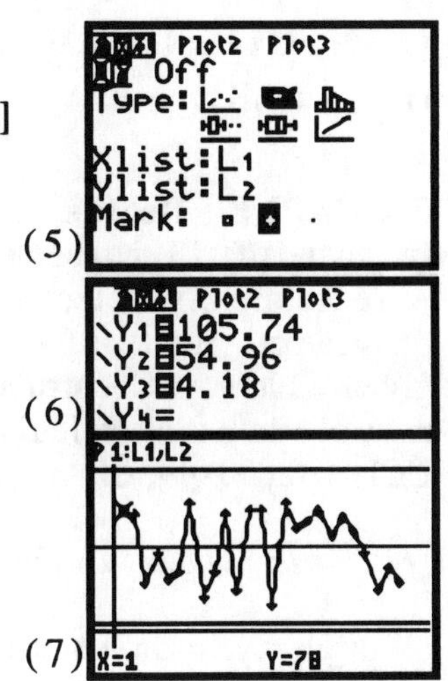


Use **STAT<CALC>1: 1-VarStats L2** for $\bar{R}$ = 54.96, from which you can calculate the control limits as in the text [pg. 660] with UCL = 105.74, CL = 54.96, and LCL = 4.18.

2. Set up Plot1 as in screen (5).
3 Set up the Y = editor with the control limits as in screen (6).
4. Press **ZOOM 9**:ZoomStat, and adjust the **WINDOW** to include the control limits by changing Ymin = -25 and Ymax = 125. Then press **TRACE** for screen (7), which is similar to the R-chart in the text [pg. 661].

## MONITORING PROCESS MEAN: CONTROL CHART FOR $\bar{x}$ [pg. 622]

**EXAMPLE Manufacturing Cola Cans** [pg. 663]: Refer to the means of the axial loads of aluminum cans (of samples of size 7 collected each day for 25 working days) given in Table 12-1 of the text and repeated below. Construct a control chart for $\bar{x}$. Use the control limits as calculated in the text [pg. 663]: UCL = 290.15, CL = 267.12, and LCL = 244.09.

252.7 247.9 270.3 267.0 281.6 269.9 257.7 272.9 273.7 259.1

275.6 262.4 256.0 277.6 264.3 260.1 254.7 278.1 259.7 269.4

266.6 270.9 281.0 271.4 277.3

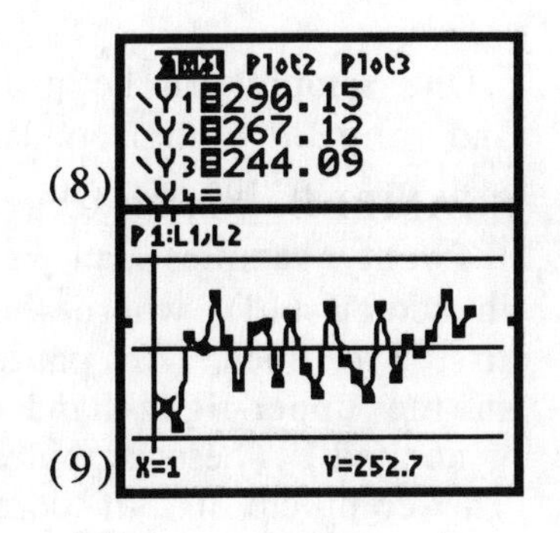


Put the day in L1 (integers from 1 to 25) and the corresponding means from above in L2. Set up Plot1 as in screen (5) and the **Y=** editor as in screen (8).

Press **ZOOM 9**:ZoomStat, and adjust the **WINDOW** to include the control limits by changing Ymin = 235 and Ymax = 300. Then press **TRACE** for screen (9), which is similar to the $\bar{x}$-chart in the text [pg. 663].

## CONTROL CHARTS FOR ATTRIBUTES (P CHART) [pg. 668]

**EXAMPLE Deaths from Infectious Diseases** [pg. 669]: In each of 13 consecutive and recent years, 100,000 subjects were randomly selected and the number who died from respiratory tract infections was recorded, with the results given below. Construct a control chart for p with the control limits as calculated in the text [pg. 670]: UCL = 0.000449, CL = 0.000288, and LCL = 0.000127.

25 24 22 25 27 30 31 30 33 32 33 32 31

Put the year in L1 (integers from 1 to 13) and the corresponding number of deaths in L2. From the Home screen, let L2÷100000 **STO▸**L2 to store the proportion of deaths in L2.

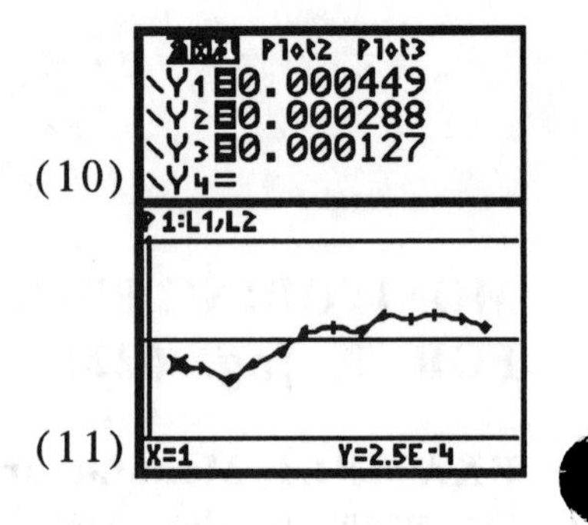

(10)

(11)

Set up Plot1 as in screen (5) and the **Y=** editor as in screen (10). Press **ZOOM 9**:ZoomStat, and adjust the **WINDOW** to include the control limits by changing Ymin = 0.00009 and Ymax = 0.0005. Then press **TRACE** for screen (11), which is similar to the p-chart in the text [pg. 670].

## TECHNOLOGY PROJECT [pg. 678]

Simulate 20 days of manufacturing heart pacemakers with a 1% rate of defective units. Use the TI-83 to simulate taking a sample of 200 pacemakers each day.

One sample can be generated as in screen (12), with rand and randBin both from **MATH**<PRB> and 2 defective as in the last line of the screen.

```
123→rand
                123
randBin(200,0.01
)
                  2
```
(12)

Twenty samples can be generated and stored in L2 as in screen (13), where the seed had been reset in the first two lines. (Be patient and notice the busy symbol in the upper-right hand corner while the list is being generated.) These samples vary from 0 to 4 defective. The complete list of defective units for the 20 samples can be seen using ▶ or looking in L2 in the Stat Editor.

```
123→rand
                123
randBin(200,0.01
,20)→L2
{2 2 0 1 1 1 4 …
```
(13)

**Note**: The text suggest another method that can be used.

# 13 Nonparametric Statistics

In this chapter you will perform most calculations using the TI-83 Plus's spreadsheet and data-sorting features. Examples are given for all the nonparametric tests in the text. When sample sizes are too large for the tables of critical values given in the text, a test statistic, z, which is normally distributed can be calculated for many methods. You can find the p-value for the calculated z by using the normalcdf function as shown in screens (17) and (32) on pages 75 and 78.

## SIGN TEST [pg. 687]

### Claims Involving Matched Pairs [pg. 689]

**EXAMPLE Reported and Measured Heights** [pg. 689]: The following data was obtained when 12 randomly selected male statistics students. Each student reported his height, then his height was measured. Use a 0.05 significance level to test the claim that there is no difference between reported height and measured height. This is the same data used on page 49.

| Subject | A | B | C | D | E | F | G | H | I | J | K | L |
|---|---|---|---|---|---|---|---|---|---|---|---|---|
| Reported height L1 | 68 | 74 | 82.25 | 66.5 | 69 | 68 | 71 | 70 | 70 | 67 | 68 | 70 |
| Measured height L2 | 66.8 | 73.9 | 74.3 | 66.1 | 67.2 | 67.9 | 69.4 | 69.9 | 68.6 | 67.9 | 67.6 | 68.8 |
| Difference L3=L1-L2 | 1.2 | 0.1 | 7.95 | 0.4 | 1.8 | 0.1 | 1.6 | 0.1 | 1.4 | -0.9 | 0.4 | 1.2 |

If male statistics students accurately report their heights, their reported heights and measured heights should be about the same, so the number of positive and negative signs should be approximately equal (six each)—but in the above table we have 11 positive signs and 1 negative sign. Are the numbers of positive and negative signs approximately equal, or are they significantly different? This is a binomial distribution with $n = 12$ and $p = 0.5$ for a mean of $n*p = 12*0.5 = 6$.

**Note**: You might want to review the binomial distribution in Chapter 4.

You want the probability in the tail or the probability of getting 1 or fewer negative differences or P(0) + P(1).

Press ↑**DISTR A:binomcdf(12,.5,1** and then **ENTER** for 0.0031738281 as in screen (1). Since this is a two-tail test, multiply this value by 2 for a p-value = 0.00635, as in the bottom line of screen (1).

```
binomcdf(12,.5,1
            .0031738281
Ans*2
            .0063476562
```
(1)

Since the p-value is less than the significance level of 0.05, we reject the null hypothesis and conclude that a majority of students report they are taller than they actually are.

## Claims Involving Nominal Data [pg. 691]

**EXAMPLE Gender Discrimination** [pg. 691]**:** A company acknowledges that about half the applicants for new jobs were men and half were women. All applicants met the basic job-qualification standards. Test the null hypothesis that men and women are hired equally by this company if of its last 100 hires only 30 were men.

$H_0$: p = 0.5 $H_1$: p ≠ 0.5

This is a binomial distribution with n = 100 and p = 0.5, and you want to find the probability of having 30 or fewer men, or P(0) + P(1) + P(2) + ... + P(29) + P(30).

```
binomcdf(100,.5,
30
     3.925069816E-5
Ans*2
     7.850139632E-5
```
(2)

Press **↑DISTR A:**binomcdf(**100,0.5,30** and then **ENTER** for 0.00003925 as in screen (2). Since this is a two-tail test, multiply this value by 2 for a p-value of 0.00007850, as in the bottom line of screen (2). Since the p-value is so small, there is sufficient evidence to warrant rejection of the claim that the hiring practices are fair.

## Claims About the Median of a Single Population [pg. 692]

**EXAMPLE Body Temperatures** [pg. 692]: Use the sign test to test the claim that the median value of the 106 body temperatures of healthy adults (used in Chapter 7) is less than 98.6°F. Thus, $H_1$: median < 98.6°F and $H_0$: median ≥ 98.6°F or the proportion of values above 98.6° F is less than 0.5). The data set has 68 subjects with temperatures below 98.6°F, 23 subjects with temperatures above 98.6°F, and 15 subjects with temperatures equal to 98.6°F. (The steps below show how you can find these values.)

Discounting the 15 temperatures equal to 98.6°F because they do not add any information to this problem, n = 68 + 23 = 91. If the median were 98.6, we would expect about half of these 91 values to be below the median (negatives) and half to be above the median (positives). This is a binomial distribution with $\mu = n*p = 91*0.5 = 45.5$ and $\sigma = \sqrt{(n*p*q)} = \sqrt{(91*(1/2)*(1/2))} = \sqrt{(91)}/2$. We want the probability of getting 23 or fewer positive values.

```
binomcdf(91,0.5,
23)
     1.261076831E-6
```
(3)

Press **↑DISTR A:**binomcdf(**91,0.5,23)** and then press **ENTER** for a p-value = 0.00000126 as in screen (3).

With such a small p-value, there is good evidence that in fact there are too few positive and significantly more negative values than would be expected. The data supports the claim that the median body temperature of healthy adults is less than 98.6°F.

If the 106 body temperatures are saved in a list (e,g., LBTEMP), the following steps show one way the numbers below and above the median given could be obtained.

1. Press **STAT 5**: SetUpEditor LBTEMP,**L1**,**L2**, **ENTER**, and then **STAT 1**:Edit for the results

in screen (4).

2 Highlight L1 at the top of screen (4), paste LBTEMP, and subtract 98.6 as in the last line. Press **ENTER** for screen (5). Since the first two values of LBTEMP are 98.6, the difference in L1 is 0. The next two values are 0.6 less than 98.6, and the fifth value is 99, or 0.4 greater than 98.6 and so on.

3 Use **STAT 2**:SortA(**L1** and then press **ENTER** to sort the differences in L1 from low negative values to high positive values.

4. Using ▼, observe that 68 values are below 98.6 (negative), as in screen (6). Zero differences go from row 69 to row 83, or 15 values are equal to 98.6. Positive differences, or values above 98.6, go from row 84 to 106, or 23 values. These results are the same as those given above.

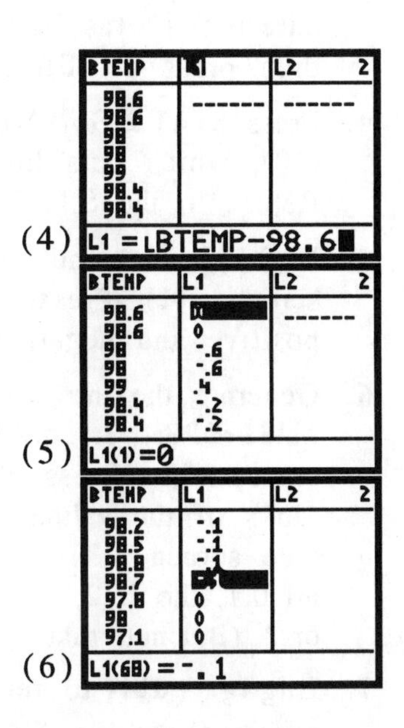

## WILCOXON SIGNED-RANKS TEST FOR MATCHED PAIRS [pg. 698]

**EXAMPLE Reported and Measured Heights** [pg. 699]: The following data (used in the first example of this chapter) was obtained when 12 students each reported his height and then his height was measured (also see page 49). Use the Wilcoxon signed-ranks test to test the claim of no difference between reported heights and measured heights. Use a significance level of α = 0.05.

| Subject | A | B | C | D | E | F | G | H | I | J | K | L |
|---|---|---|---|---|---|---|---|---|---|---|---|---|
| Reported height L1 | 68 | 74 | 82.25 | 66.5 | 69 | 68 | 71 | 70 | 70 | 67 | 68 | 70 |
| Measured height L2 | 66.8 | 73.9 | 74.3 | 66.1 | 67.2 | 67.9 | 69.4 | 69.9 | 68.6 | 67.9 | 67.6 | 68.8 |
| Difference L3=L1-L2 | 1.2 | 0.1 | 7.95 | 0.4 | 1.8 | 0.1 | 1.6 | 0.1 | 1.4 | -0.9 | 0.4 | 1.2 |

The following steps show a way of finding the sum of the positive ranks and the sum of the absolute values of the negative ranks.

1. Under **STAT 1**:Edit... put the reported heights in L1 and the measured heights in L2. Highlight L3 and then enter L1-L2 on the bottom line. Press **ENTER** to calculate the difference, as in screen (7).

2. Highlight L4 and enter abs( L3, as in the bottom line of screen (7). Press **ENTER** for screen (8) with all positive values in L4 which is made clearer in screen (9) with the 10th row being positive in L4 but negative in L3. **Note**: abs( from **MATH**<NUM>**1**.

3. Copy the differences in L3 into L5 by highlighting L5,

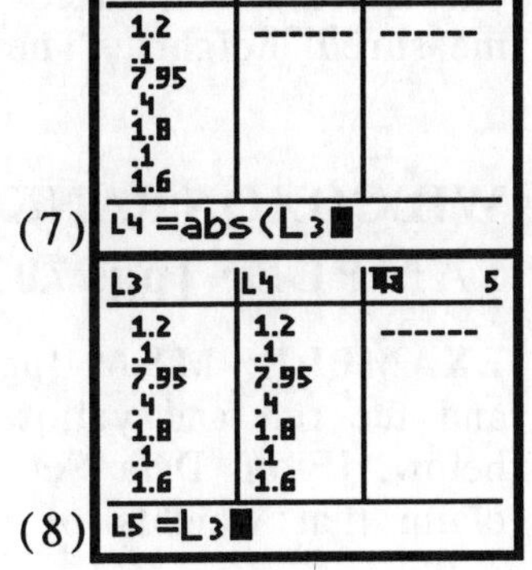

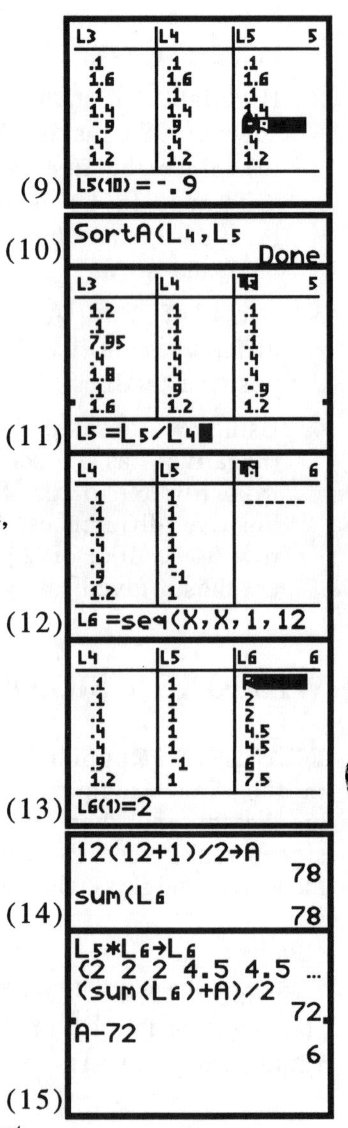


entering L3 (as in the bottom line of screen (8)), and then press **ENTER** for screen (9).

4. Press **STAT 2**:SortA(L4,L5 **ENTER**, for Done as in screen (10), which puts L4 in order and carries along L5. Return to the STAT editor for screen (11).
5. Highlight L5 and enter L5 ∍ L4, as in the last line of screen (11). Pressing **ENTER** gives you a column of positive and negative 1 s in L5, as in screen (12).
6. Generate the integers from 1 to 12 in L6 with **↑LIST<OPS> 5:seq(X,X,1,12** as in the bottom line of screen (12). Press **ENTER** and modify L6 so that the ranks of the values in each row of L4 are given in L6 (see screen (13)). Since the first three values in L4 are all 0.1, the 1, 2, and 3 in L6 are changed to their average, or 2. (But note that 2 + 2 + 2 = 1 + 2 + 3.)
7. Engage **↑QUIT** to return to the Home screen. Type 12(12+1)∍2 **STO▸ A ENTER** for the sum of the ranks (or 78) to be stored as A. Engage **↑LIST<MATH>5**:sum( **L6** for the sum of the values in L6, which should also be 78. If you get a different answer, check your rankings (see screen (14)).
8. Now have L5 *L6 **STO▸** L6 **ENTER**, which puts a sign on the ranks in L6. In this example, the first five are positive, and the sixth is negative; the rest are positive.
9. (sum(L6) +A)∍2 **ENTER** for 72 or the sum of the ranks for the positive differences. Type **A - ↑ANS ENTER** for 6, the sum of the ranks for the negative differences. Of course, this problem has just one such value.

Since 6 is the smaller of the two and n = 12 | 30, we find that the critical value in text Table A-8 is 14; and since the test statistic of 46 is less than 14, we reject the null hypothesis. It appears that there is a difference between reported and measured heights. The majority of students exaggerate their height.

## WILCOXON RANK-SUM TEST FOR TWO INDEPENDENT SAMPLES [pg. 703]

**EXAMPLE M&M** [pg. 705]: Samples of M&M plain candies are randomly selected, and the red and yellow M&Ms are weighed, with the results listed in the table below. [From Data Set 10 in Appendix B.] At the 0.05 level of significance, test the claim that weights of red M&Ms and yellow M&Ms have the same distribution.

The text bases its results on the sum of the ranks of the red M&Ms, which is R = 469.5. You will see how to calculate this in the numbered steps below. Since $n_1 = 21$ and $n_2 = 26$ are both greater than 10, you can use a normal distribution with $\mu_r = n_1(n_1 + n_2 + 1)/2$

Or type **21(21+26+1)/2 STO▶ M** and then press **ENTER** for 504, as in the top lines of screen (16).

$\sigma_r = \sqrt{(n_1 * n_2 * (n_1 + n_2 + 1)/12)}$, or type

**√(21*26*(21+26+1)/12 STO▶ S**

and then press **ENTER** for 46.7333.

$z = (R \cong \mu_r)/\sigma_r = (469.5 \cong 504)/46.733 = -0.7382.$

```
21(21+26+1)/2→M
             504
√(21*26*(21+26+1
)/12)→S
     46.73328578
(469.5-M)/S
    -.7382318496
```

(16)

Find the area in the left tail of the normal distribution using **↑DISTR 2:normalcdf(⁻E99,⁻.738.** Then press **ENTER** for a p-value = 0.230257*2 = 0.4605, as in screen (17). Thus we fail to reject the null hypothesis of similar distributions because the p-value is so large.

```
normalcdf(⁻E99,⁻
.738
    .2302571507
Ans*2
    .4605143014
```

(17)

**Note:** The mean rank for red M&Ms is 469.5/21 = 22.36, and the mean rank for yellow M&Ms is 658.5/26 = 25.33. These means are fairly close, so we are not surprised that we fail to reject the null hypothesis.

| L1 | L2 |
|---|---|
| 0.87 | 1 |
| 0.933 | 1 |
| 0.952 | 1 |
| 0.908 | 1 |
| 0.911 | 1 |
| 0.908 | 1 |
| 0.913 | 1 |
| 0.983 | 1 |
| 0.92 | 1 |
| 0.936 | 1 |
| 0.891 | 1 |
| 0.924 | 1 |
| 0.874 | 1 |
| 0.908 | 1 |
| 0.924 | 1 |
| 0.897 | 1 |
| 0.912 | 1 |
| 0.888 | 1 |
| 0.872 | 1 |
| 0.898 | 1 |
| 0.882 | 1 |
| 0.906 | 0 |
| 0.978 | 0 |
| 0.926 | 0 |
| 0.868 | 0 |
| 0.876 | 0 |
| 0.968 | 0 |
| 0.921 | 0 |
| 0.893 | 0 |
| 0.939 | 0 |
| 0.886 | 0 |
| 0.924 | 0 |
| 0.91 | 0 |
| 0.877 | 0 |
| 0.879 | 0 |
| 0.941 | 0 |
| 0.879 | 0 |
| 0.94 | 0 |
| 0.96 | 0 |
| 0.989 | 0 |
| 0.9 | 0 |
| 0.917 | 0 |
| 0.911 | 0 |
| 0.892 | 0 |
| 0.886 | 0 |
| 0.949 | 0 |
| 0.934 | 0 |

The following steps give a method for finding the needed ranks.

1. The weights, in grams, are given in the first column of the table on the right. Put these values in L1. The 21 weights of the red M&Ms are listed first and are identified by a 1 in the second column. The other 26 weights are yellow M&Ms and are identified by a 0. Put 21 1 s followed by 26 0 s in L2.
2. Make a copy of L1 into L3. Make a copy of L2 into L5.
3. Press **STAT 2**:SortA(L3, L5 **ENTER** to sort the values in L3 and carry along the values in L5.
4. Generate the integers from 1 to 47 in L4 with **↑LIST<OPS> 5:seq(X,X,1,47,1** and then modify L4 so that it has the ranks of the values in L3 in L4 ( screen (18)).
   **Note**: Notice how the tie in the 7th and 8th slots was handled (both given a rank of 7.5) in screen (18). The 10th and 11th values are also the same and get the ranks of 10.5. The 20th, 21st, and 22nd items, which are the same, all get a rank of 21. The 24th and 25th values both get 24.5; the 31st, 32nd, and 33rd values all get ranked 32.

| L3 | L4 | L5 |
|---|---|---|
| .872 | 3 | 1 |
| .874 | 4 | 1 |
| .876 | 5 | 0 |
| .877 | 6 | 0 |
| .879 | 7.5 | 0 |
| .879 | 7.5 | 0 |
| .882 | | 1 |

L4(9) =9

(18)

5. After entering all the ranks, press **↑QUIT** to return to the Home screen. Type 47(47+1)∋2 **STO▶ A** (for 1128, the sum of the integers from 1 to 47 stored in **A**). Then find the sum(L4) as a check on your work (see screen (19)).

6. Multiply the 1 s and 0 s in L5 by the ranks in L4 and put the results in L6 (L4* L5 **STO▸** L6).
   **Note:** This will copy all the ranks for the red M&M sample, but the yellow M&M ranks will be zeroed out.
7. Then sum(L6 **ENTER** gives the sum of the red M&M ranks, or 469.5 = $R_1$ of the text, as shown in screen (20). Subtracting 469.5 from the total sum of all the ranks gives us the sum of the ranks for the yellow M&Ms, or A≅↑**ANS ENTER** for 658.5 = $R_2$ of the text.

```
47(47+1)/2→A
                1128
sum(L4)
                1128
```
(19)

```
L4*L5→L6
{0 2 3 4 0 0 0 ...
sum(L6
               469.5
A-Ans
               658.5
```
(20)

## KRUSKAL–WALLIS TEST [pg. 710]

**EXAMPLE Crash Test Dummies** [pg. 712 Chapter Problem for Chapter 11]: The data in the table at the right are the head injury measurements, in L1, taken from crash test dummies for four weight categories of cars designated by 1, 2, 3, or 4 in L2. Using the head injury measurements, is there sufficient evidence to conclude that head injuries for the four weight categories are not all the same?

| L1 | L2 |
|---|---|
| 681 | 1 |
| 428 | 1 |
| 917 | 1 |
| 898 | 1 |
| 420 | 1 |
| 643 | 2 |
| 655 | 2 |
| 442 | 2 |
| 514 | 2 |
| 525 | 2 |
| 469 | 3 |
| 727 | 3 |
| 525 | 3 |
| 454 | 3 |
| 259 | 3 |
| 384 | 4 |
| 656 | 4 |
| 602 | 4 |
| 687 | 4 |
| 360 | 4 |

The sum of the ranks of head injury measurements for each car weight category are shown in the text and are calculated in the numbered steps that follow. They are $R_1 = 64$ for the lightest cars, $R_2 = 52.5$, $R_3 = 44.5$, and $R_4 = 49$ for the heaviest cars . Since $n_1 = n_2 = n_3 = n_4 = 5$ are each at least 5, H (that follows) is chi-square distributed with k ≅ 1 = 4 ≅ 1 = 3 degrees of freedom. Below, $N = n_1 + n_2 + n_3 + n_4 = 5 + 5 + 5 + 5 = 20$.

$H = 12/(N(N+1))(R_1^2/n_1 + R_2^2/n_2 + R_3^2/n_3 + R_4^2/n_4) \cong 3(N + 1)$

$= 12/(20*21)*(64^2/5 + 52.5^2/5 + 44.5^2/5 + 49^2/5) \cong 3*21$

$= 1.1914$

Press ↑**DISTR** 7:χ²cdf(**1.1914,E99,3)** and then **ENTER** for the p-value = 0.7551 (see screen (21)). Since the p-value is greater than 0.05, we fail to reject the null hypothesis of identical populations. On the basis of the available evidence, the four weight categories appear to have the same distribution of head injury measurements.

```
χ²cdf(1.1914,E99
,3)
          .7550676774
```
(21)

Use the method described in the following steps to find the ranks of each of the four samples.

1. With the data in L1 and the weights coded from 1 to 4 in L2 as in the table on page 88, make a copy of L1 in L3 (with L1 **STO▸** L3 on the Home screen).
2. Press **STAT 2**:SortA(L3, L2 **ENTER** to sort the values in L3 and carry along the values in L2 (see L2 and L3 in screen(22) but notice that the cursor was moved to the 10th row).

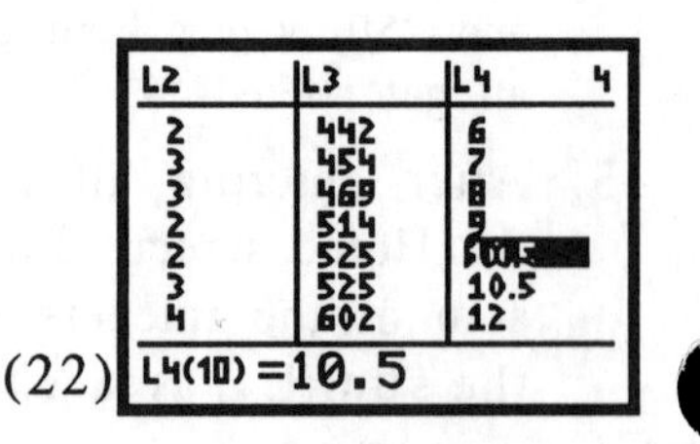

(22)

3. Generate the integers from 1 to 20 in L4 with **↑LIST<OPS> 5:seq(X,X,1,20 STO▸** L4 and then modify L4 so that it has the ranks of the values in L3 in L4 (see screen (22)). The 10th and 11th values are both 525, so they are both given the same rank, 10.5 as 10 + 11 = 10.5 +10.5 =21. There are no other values that are the same.
4. Screen (23) shows that the sum of the integers from 1 to 20 is 210. As a check, the sum of the ranks in L4 is also 210.
5. SortA(L2, L3, L4 to sort the values in L2, putting all the 1 s first, followed by the 2 s, and so on, and carry along the values in L3 and L4.
6. Generate 20 zeros in L5 with **↑LIST<OPS> 5:seq(0,X,1,20 STO▸** L5 and then replace the first 5 values with 1 s, as shown in screen (24), with all other values 0 s.
7. Multiply L4 by L5 and store the results in L6. Only the first 5 values, or the ranks that go with the lightest cars, will be in L6; all other rows will record a zero so that when you **sum(L6** you get R1 = 64 (see screen (25)).

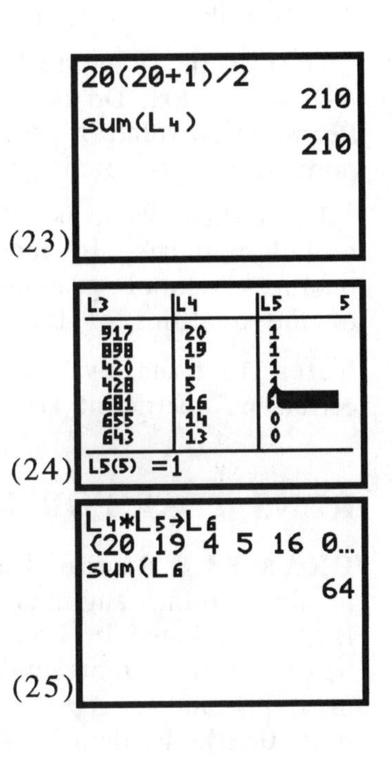


Repeat the above two steps, but with 1 next to the 6th through 10th values and all other values zero and so on.

## RANK CORRELATION [pg. 719]

**EXAMPLE Business School Rankings** [pg. 722]: *Business Week* magazine ranked business schools two different ways. Corporate rankings were based on surveys of corporate recruiters, and graduate rankings were based on surveys of MBA graduates. The table at the right is based on the results for 10 schools. Is there a correlation between the corporate rankings and the graduate rankings? The linear correlation coefficient r (section 9-2) should not be used because it requires normal distributions and the data consists of ranks, which are not normally distributed. Instead, the rank correlation coefficient should be used to test the claim that there is a relationship between corporate and graduate rankings (that is, $\rho_s$ 0). Use $\alpha = 0.05$.

| School | Corp. Rank(L1) | Grad. Rank(L2) |
|---|---|---|
| PA | 1 | 3 |
| NW | 2 | 5 |
| Chi | 4 | 4 |
| Sfd | 5 | 1 |
| Hvd | 3 | 10 |
| MI | 6 | 7 |
| IN | 8 | 6 |
| Clb | 7 | 8 |
| UCLA | 10 | 2 |
| MIT | 9 | 9 |

With the data in L1 and L2, the scatter diagram in screen (26) shows little evidence of a relationship between the two variables.

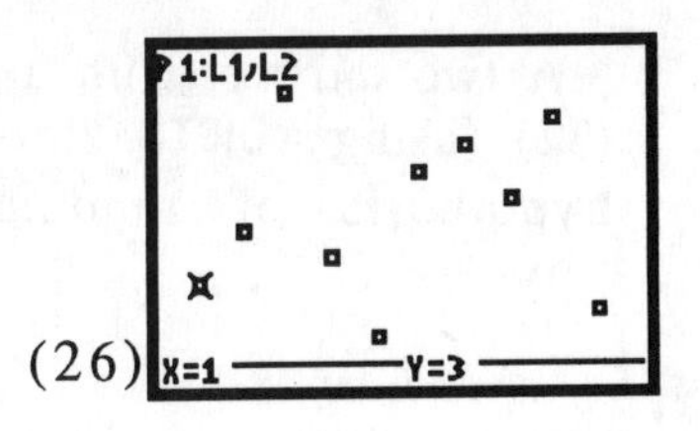

(26)

Press **STAT<TESTS>E**:LinRegTTest and set up the screen for screen (27).

Highlight Calculate in the bottom line of screen (27), and then press **ENTER** for screens (28) and (29). The last line of screen (29) gives $r_s = 0.1030$.

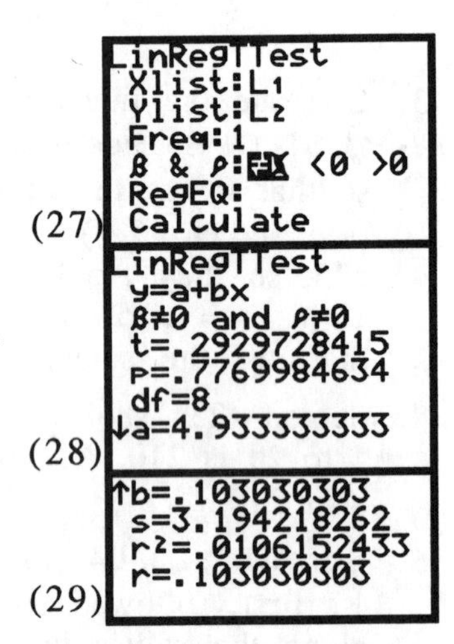


To test the hypothesis, use the critical value from Table A-9 of the text. Do not use Table A-6 (which was used for the Pearson correlation coefficient) because it requires the populations sampled to be normally distributed.

The critical value is 0.648 > 0.103, so we fail to reject the null hypothesis. It appears that corporate recruiters and business school graduates have different perceptions of the qualities of the schools.

**Note**: The t and p-values of screen (28) are for the Pearson correlation coefficient and do not apply for the ranks.

## RUNS TEST FOR RANDOMNESS [pg. 729]

**EXAMPLE Boston Rainfall On Mondays** [pg. 733]: Refer to the rainfall amounts from 52 consecutive Mondays in Boston as listed in Data Set 17 [Appendix B] and repeated below with 0 representing no rain (indicated by a value of 0.00) and 1 representing some rain (any value greater than 0.00). Is there sufficient evidence to support the claim that rain on Mondays is not random?

0000 1 0 1 00 1 00 1 000 1 00 111 0000

1 0 1 0 111 0 1 000 1 000 1 0 1 00 1 000 1

$n_1 = 33$ The number of zeros or dry days

$n_2 = 19$ The number of ones or days of some rain

$G = 30$ The number of runs or groupings above

Since $n_1 > 20$ we can use formulas 13-2 and 13-3 [pg. 732] for $\mu_G = 25.115$ and $\sigma_G = 3.306$. Both use only $n_1$ and $n_1$ as given in screens (30) and (31).

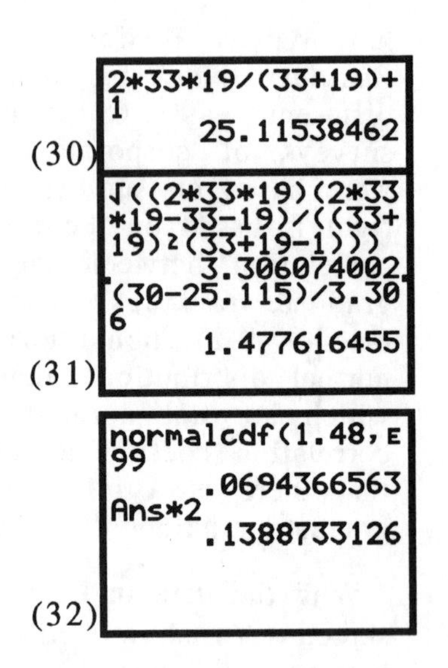


The test statistic z = 1.48 follows as below and in the bottom lines of screen (31).

$$z = (G - \mu_G)/\sigma_G = (30 - 25.115)/3.306 = 1.48$$

A two tail test with a p-value = 0.1389 > 0.05 as in screen (32) (using **↑DISTR 2**:normalcdf). We fail to reject the null hypothesis of randomness.

# Appendix

## Loading the Data Apps, Data, Groups, and Programs from a Computer or Another TI-83 Plus.

This appendix contains information on transferring data and programs from the CD-ROM that came with your main text and available from the publisher. Also given are examples of installing a data set from the data Apps and how to make and install your own group.

The CD-ROM contains the data sets from Appendix B of *Elementary Statistics* (8th Ed.) by Mario F. Triola and are given in an Apps (or application called TRIOLA8E APP) and as individual lists in ASCII format as text files (with extensions of .txt) for TI-83s. Two programs, one named A1ANOVA.83p (used in Chapter 11) and one named A2MULREG.83p (used in the later parts of Chapter 9) are also on the CD-ROM.

Your instructor will probably load the data (and programs if needed) into your TI-83 Plus or you can transfer them from the CD-ROM with your computer if you have the TI-GRAPH LINK software and cable available from Texas Instruments. See the guidebook that comes with your TI-83 Plus for information on where you can find assistance with the TI-GRAPH LINK.

## Loading Data or Programs from One TI-83 Plus to Another

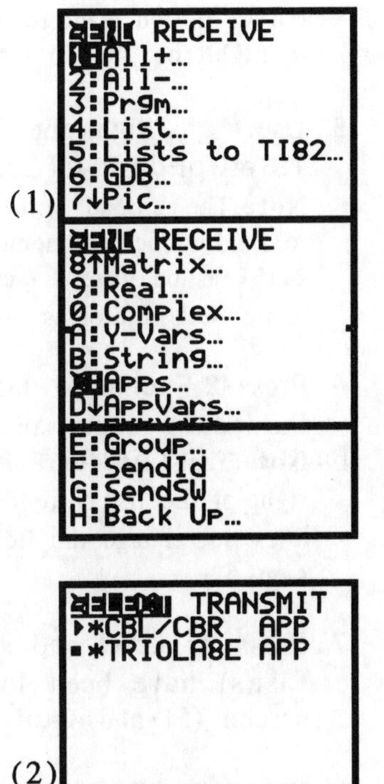


1. Connect one TI-83 Plus to another with the cable that came with the calculator. The TI-83 Plus link port is located at the center of the bottom edge of the calculator.
2. Press ↑LINK on the TI-83 Plus that is to receive the data or program (for screen (1)), and then press ▶ to highlight RECEIVE. Press 1 or ENTER to select Receive and have the message "Waiting..." displayed.
3. Press ↑LINK on the TI-83 Plus that is to send the program or data, and then press 3 for Program or 4 for List or E for Group or in this example C for Apps and a screen like screen (2). Use ▼ to point out TRIOLA8E APP and then press ENTER. A small darkened box will appear in front of the name (easier to see if you move the cursor pointer up.).
   **Note**: If this was a list or program you could repeat the process for all lists or programs to be sent.
4. Press ▶ to highlight TRANSMIT. Press 1 or ENTER to transmit the App, lists, programs, groups etc.
   **Note**: If transmitting an APP the receiving calculator will signal "garbage collecting" then "receiving", and then "validating" the APP before indication "Done" so be patient.

**Note**: If the name is already in use on the receiving calculator you will get a screen like (3) which indicates that the list CLASS that is being transmitted has a list by that name already on the receiving calculator. Pressing 2:Overwrite saves the desired data as CLASS.

## Using the Data App (for TI-83 Plus only)

1. Press **APPS** at B4 for a screen like screen (4).
   **Note**: We assume the TRIOLA8E APP has been installed. The number of applications could differ from screen (4).

2. Use the ▼ key to highlight the number in front of TRIOLA8E. Press **ENTER** for a title screen that soon changes to screen (5) with the names of the data sets in Appendix B from Triola's 8th Edition.

3. Press **3**:DIAMONDS for screen (6) with the names of the lists in data set diamonds.

4. Use the ▼ key to move the pointer (next to TABLE in screen (6) ) to a desired list and press **ENTER** to Select the list by having a small black square designate the list.
   **Note**: If you want to load all the Lists in the data set use ▶ to highlight 'All' to make that selection.

5. Use ▶ to highlight Load as in screen (7).Press **1**:SetUpEditor for screen (8)
   **Note**:The 2:Load option loads the lists from archive memory to random access memory and can be accessed by pressing ↑**LIST** as on page 8 (screen (24)). The way to reload the editor using SetUpEditor is on page 9 and 10.

6. Press **2**:Exchange Lists to have these lists loaded into the editor alone and for screen (9).
   **Note**:The 1:Add to Editor option adds the list to be loaded to the right of the list currently in the editor. The 3: No Change option does not change the Editor but loads the data as in the Note of step 5.

7. Press **2**: Load and screen (10) that informs that the List(s) have been loaded. Pressing any key returns to screen (5) and Quit returns to the Home screen.

8. Pressing **STAT** 1:Edit for the current example gives the three selected list in the editor as in screen (11)

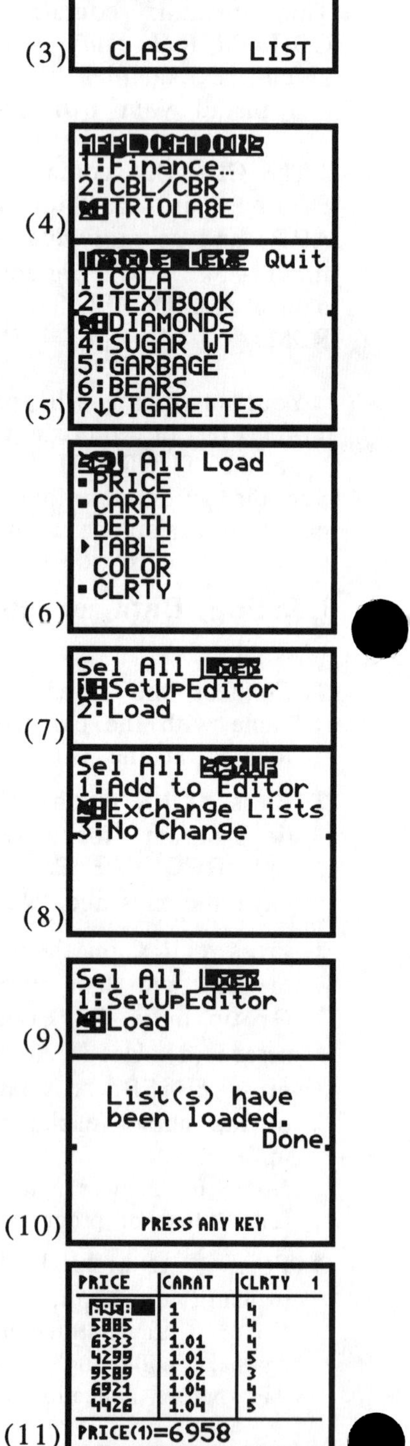

## Grouping and Ungrouping (for TI-83 Plus only)

Just as the Data Apps has lists of the data sets in the Appendix of the main text grouped together you may want to group some lists and /or matrices and/or programs etc. The advantage of groups is that they are saved in Archive memory and do not take up room in RAM memory until you want to use them.

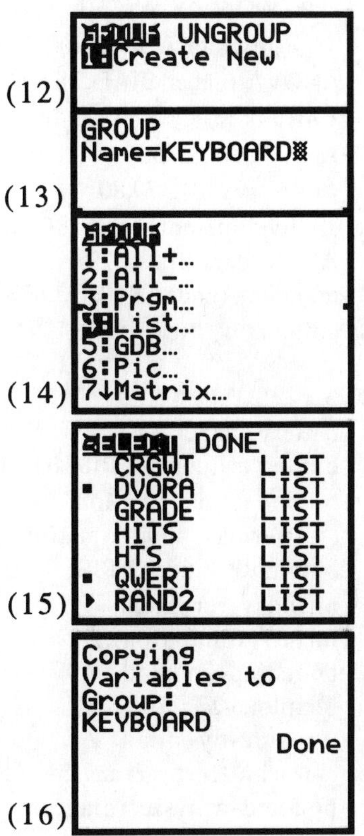


**Example** Group the Dvorax data saved as DVORA (page 23) and Qwerty data saved as QWERT (page 15) into a group called KEYBOARD.

1. Press ↑**MEM** (above the + key) then **8**:Group for screen (12). Press **ENTER** for screen (13) after the group name KEYBOARD has been typed.

2. Press **ENTER** for screen (14) and then press **4**:List for a screen like screen (15). Select the lists by using the ▼ key to line up the arrow and press **ENTER** for the little black squares in the left margin.

3 Press ▶ to highlight DONE then press **ENTER** for screen (16).

Now if you deleted the two lists from RAM, as explained on page 7, they would only be in Archive memory. To reinstall them to RAM return to screen (12) and highlight UNGROUP then press **ENTER** and select group KEYBOARD to ungroup. Even though the lists will again reside in RAM it also remains in Archive memory unless deleted as a Group similar to what was done on page 7 for lists.

# Index

# TI-83 Plus Quick Reference

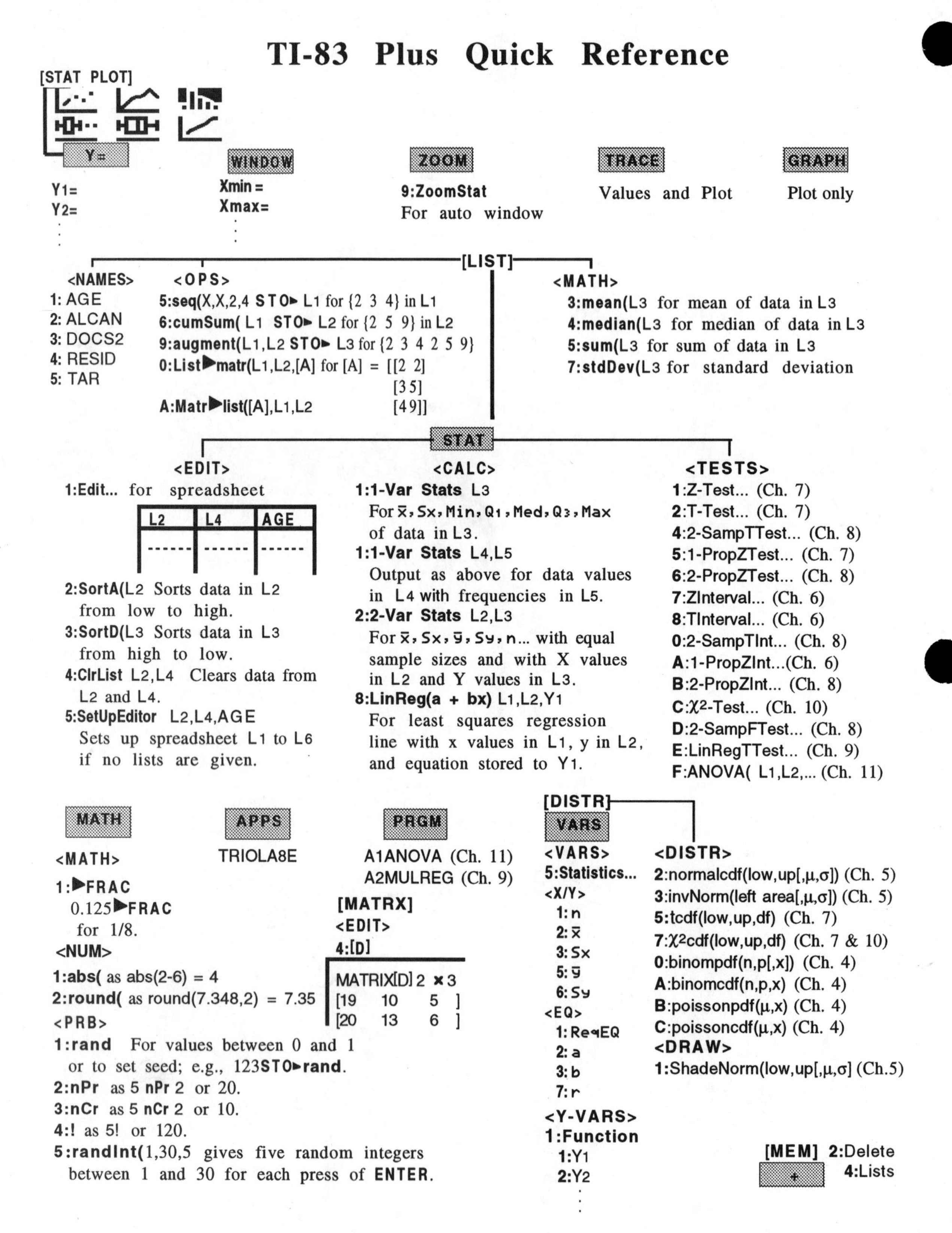